ELEMENTARY

GEOMETRY,

WITH

APPLICATIONS IN MENSURATION.

BY CHARLES DAVIES, L.L. D.

AUTHOR OF FIRST LESSONS IN ARITHMETIC, ELEMENTARY ALGEBRA, PRACTICAL GEOMETRY, ELEMENTS OF SURVEYING, ELEMENTS OF DESCRIPTIVE GEOMETRY, SHADES, SHADOWS AND PERSPECTIVE, ANALYTICAL GEOMETRY, DIFFERENTIAL AND INTEGRAL CALCULUS.

NEW YORK:
PUBLISHED BY A. S. BARNES & CO.,
No. 51 JOHN-STREET.
CINCINNATI:—H. W. DERBY & CO.
1850.

F. C. GUTIERREZ, PRINTER,
CORNER OF JOHN AND DUTCH STREETS,
NEW YORK.

PREFACE.

Those who are conversant with the preparation of elementary text-books, have experienced the difficulty of adapting them to the various wants which they are intended to supply.

The institutions of education are of all grades, from the college to the district school, and although there is a wide difference between the extremes, the level, in passing from one grade to the other, is scarcely broken.

Each of these classes of seminaries requires text-books adapted to its own peculiar wants; and if each held its proper place in its own class, the task of supplying suitable works would not be difficult.

An indifferent college is generally inferior in the system and scope of its instruction to the academy or high school; while the district school is often found to be superior to its neighboring academy.

The Geometry of Legendre, embracing a complete course of Geometrical science, is all that is desired in the colleges and higher seminaries; while the Practical Geometry, published a few years since, meets the wants of those schools which are strictly elementary in their systems of instruction.

But still a large class of seminaries remained unsupplied with a suitable text-book on Geometry: viz., those where the pupils are carried beyond the acquisition of facts and mere practical knowledge, but have not time to go through with a full course of mathematical studies

It is for such, that the following work is designed. It has been the aim of the author to present the striking and important truths of Geometry in a form more simple and concise than could be adopted in a complete treatise, and yet to preserve the exactness of rigorous reasoning.

In this system of Geometry nothing has been taken for granted, and nothing passed over without being fully demonstrated.

In order, however, to render the applications of Geometry to the mensuration of surfaces and solids complete in itself, a few rules have been given which are not demonstrated. This forms an exception to the general plan of the work, but being added in the form of an appendix, it does not materially break its unity.

That the work may be useful in advancing the interests of education, is the hope and ardent wish of the author.

HARTFORD,

April, 1841.

CONTENTS.

APPLICATIONS OF GEOMETRY.

ELEMENTARY

GEOMETRY.

BOOK I.

DEFINITIONS AND REMARKS.

1. A *Line* is length without breadth or thickness.

2. The *Extremities of a Line* are called points: and any place between the extremities is also called a point.

3. A *Straight Line* is the shortest distance between two points. Thus *AB* is a straight line, and is the shortest distance from *A* to *B*.

A————B

4. A *Curve Line* is one which changes its direction at every point. Thus, *ABC* is a curve line.

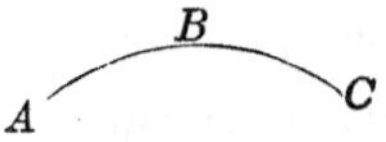

A B C

5. The word *Line*, used by itself, means a straight line; and the word *Curve*, means a curve line.

6. A *Surface* is that which has length and breadth, with out height or thickness.

7. A *Plane Surface* is that which lies even throughout its whole extent, and with which a straight line, laid in any direction, will exactly coincide in its whole length.

8. A *Curved Surface* has length and breadth without thickness, and like a curve line is constantly changing its direction.

9. A *Solid* or *Body* is that which has length, breadth, and thickness. Length, breadth, and thickness, are called dimen-

sions. Hence, a solid has three dimensions, a surface two, and a line one. A point has no dimensions, but position only.

10. *Geometry* treats of lines, surfaces, and solids.

11. A *Demonstration* is a course of reasoning which establishes a truth.

12. An *Hypothesis* is a supposition on which a demonstration may be founded.

13. A *Theorem* is something to be proved by demonstration

14. A *Problem* is something proposed to be done.

15. A *Proposition* is something proposed either to be done or demonstrated—and may be either a problem or a theorem.

16. A *Corollary* is an obvious consequence, deduced from something that has gone before.

17. A *Scholium* is a remark on one or more preceding propositions.

18. An *Axiom* is a self evident proposition.

OF ANGLES.

19. An *Angle* is the opening or inclination of two lines which meet each other at a point.

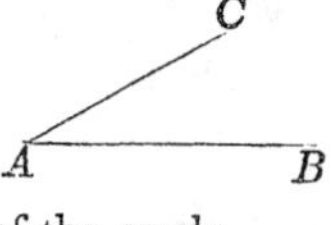

Thus, the lines AC, AB, form an angle at the point A. The lines AC, AB are called the *sides* of the angle ; and the point A, at which they meet, is called the vertex of the angle.

An angle is generally read, by placing the letter at the vertex in the middle. Thus, we say, the angle CAB. We may, however, say simply, the angle A.

20. One line is said to be perpendicular to another when it inclines no more to the one side than to the other.

The two angles formed are then equal to each other. Thus, if the line DB is perpendicular to AC, the angle DBA will be equal to DBC.

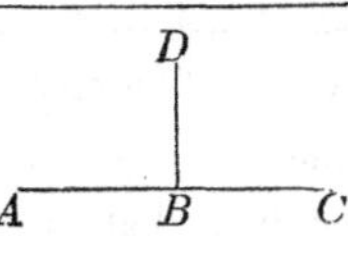

21. When two lines are perpendicular to each other, the angles which they form are called right angles. Thus, DBA and DBC are called right angles.

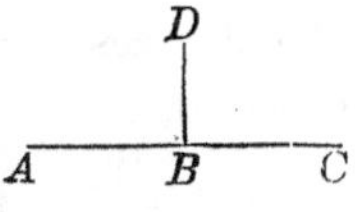

22. An acute angle is less than a right angle. Thus, DBC is an acute angle.

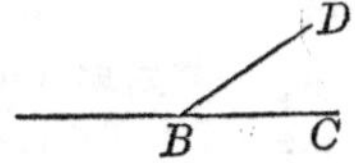

23. An obtuse angle is greater than a right angle. Thus, DBC is an obtuse angle.

24. The circumference of a circle is a curve line all the points of which are equally distant from a certain point within called the centre.

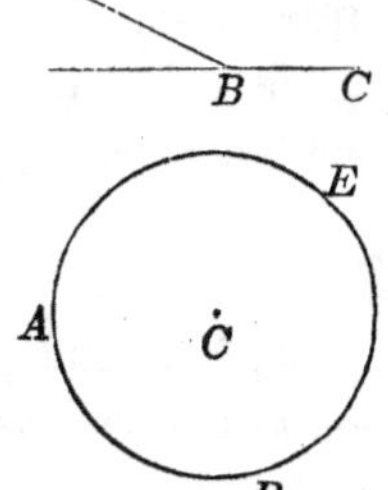

Thus, if all the points of the curve AEB are equally distant from the centre C, this curve will be the circumference of a circle.

25. Any portion of the circumference, as AED, is called an *arc*.

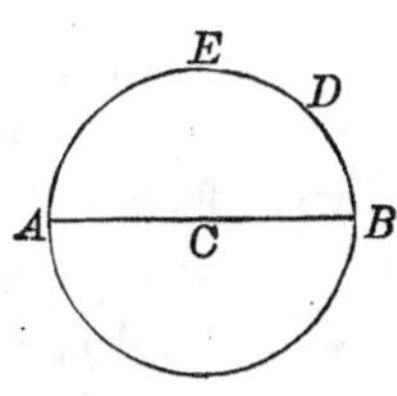

26. The diameter of a circle is a straight line passing through the centre and terminating at the circumference. Thus, ACB is a diameter.

27. One half of the circumference, as ACB is called a *semicircumference;* and one quarter of the circumference, as AC, is called a *quadrant*.

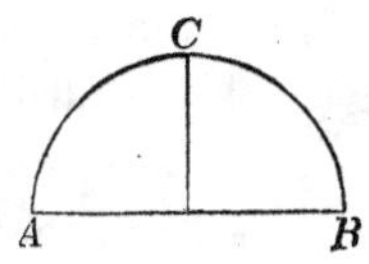

28. The circumference of a circle is used for the measurement of angles. For this purpose it is divided into 360 equal parts called degrees, each degree into 60 equal parts called minutes, and each minute into 60 equal parts called seconds. The degrees, minutes, and seconds are marked thus $^\circ$ $'$ $''$; and $9^\circ\ 18'\ 16''$, are read, 9 degrees 18 minutes and 16 seconds.

29. Let us suppose the circumference of a circle to be divided into 360 degrees, beginning at the point B. If through the point of division marked 40, we draw CE, then, the angle ECB will be equal to 40 degrees. If CF were drawn through the point of division marked 80, the angle BCF would be equal to 80 degrees.

OF LINES.

30. Two straight lines are said to be *parallel*, when being produced either way, as far as we please, they will not meet each other.

31. Two curves are said to be parallel or *concentric*, when they are the same distance from each other at every point.

32. Oblique lines are those which approach each other, and meet if sufficiently produced.

33. Lines which are parallel to the horizon, or to the water level, are called horizontal lines.

34. Lines which are perpendicular to the horizon, or to the water level, are called vertical lines.

OF PLANE FIGURES.

35. A Plane Figure is a portion of a plane terminated on all sides by lines, either straight or curved.

36. If the lines which bound a figure are straight, the space which they inclose is called a *rectilineal* figure, or *polygon.* The lines themselves, taken together, are called the *perimeter* of the polygon. Hence, the perimeter of a polygon is the sum of all its sides.

37. A polygon of three sides is called a triangle.

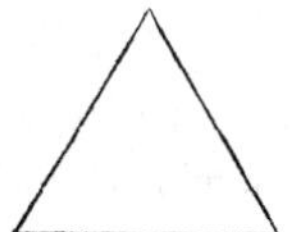

38. A polygon of four sides is called a quadrilateral.

39. A polygon of five sides is called a pentagon.

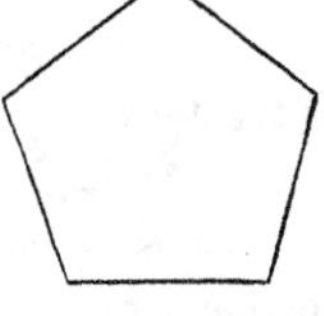

40. A polygon of six sides is called a hexagon.

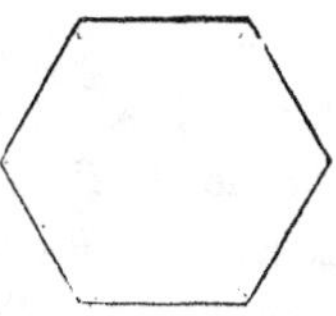

41. A polygon of seven sides is called a heptagon.

42. A polygon of eight sides is called an octagon.

43. A polygon of nine sides is called a nonagon.

44. A polygon of ten sides is called a decagon.

45. A polygon of twelve sides is called a dodecagon.

46. There are several kinds of triangles.

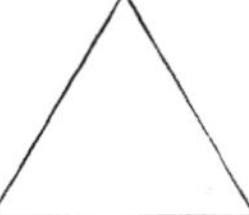

First. An equilateral triangle, which has its three sides all equal.

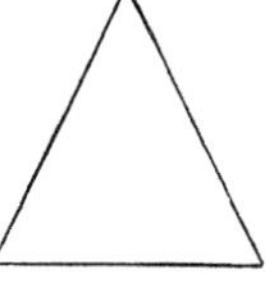

Second. An isosceles triangle, which has two of its sides equal.

Third. A scalene triangle, which has its three sides all unequal.

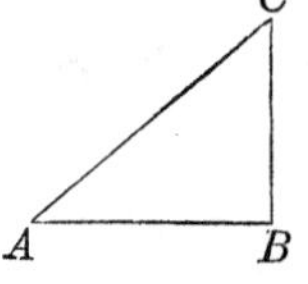

Fourth. A right angled triangle, which has one right angle.

In the right angled triangle *ABC*, the side *AC*, opposite the right angle, is called the hypothenuse.

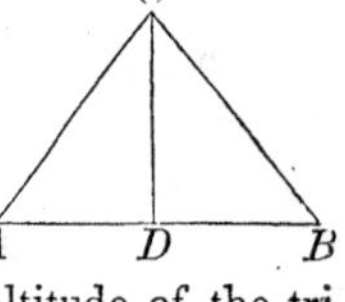

47. The base of a triangle is the side on which it stands. Thus, *AB* is the base of the triangle *ACB*.

The altitude of a triangle is a line drawn from the angle opposite the base and perpendicular to the base. Thus, *CD* is the altitude of the triangle *ACB*.

48. There are several kinds of quadrilaterals.

First. The square, which has all its sides equal, and all its angles right angles.

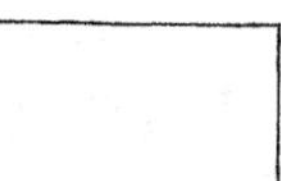

Second. The rectangle, the opposite sides of which are parallel and its angles right angles.

Third. The parallelogram, which has its opposite sides parallel, but its angles not right angles.

Fourth. The rhombus, which has all its sides equal, and the opposite sides parallel, without having its angles right angles.

Fifth. The trapezoid, which has only two of its sides parallel.

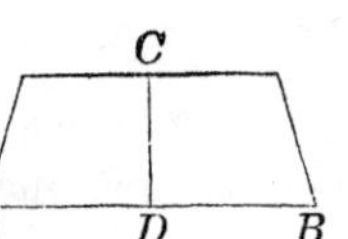

49. The base of a figure is the side on which it stands, and the altitude is a line drawn from the opposite side, or angle, perpendicular to the base. Thus, AB is the base and CD is the altitude of the trapezoid.

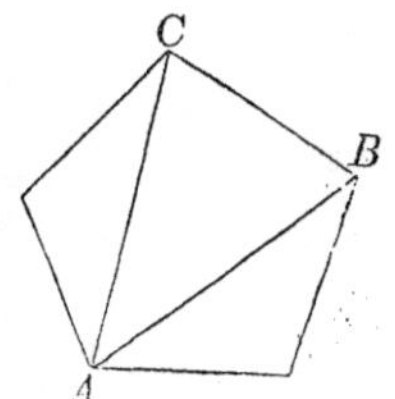

50. A diagonal is a line joining the vertices of two angles not adjacent. Thus, AB and AC are diagonals.

AXIOMS.

1. Things which are equal to the same thing are equal to each other.

2. If equals be added to equals, the wholes will be equal.

3. If equals be taken from equals, the remainders will be equal.

4. If equals be added to unequals, the wholes will be unequal.

5. If equals be taken from unequals, the remainders will be unequal.

6. Things which are double of equal things, are equal to each other.

7. Things which are halves of the same thing, are equal to each other.

8. The whole is greater than any of its parts.

9. The whole is equal to the sum of all its parts.

10. All right angles are equal to each other.

11. Magnitudes, which being applied to each other, coincide throughout their whole extent, are equal.

PROPERTIES OF POLYGONS.

THEOREM I.

Every diameter of a circle divides the circumference into two equal parts.

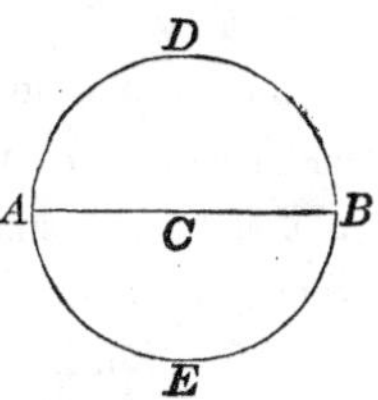

Let *ADBE* be the circumference of a circle, and *ACB* a diameter: then will the part *ADB* be equal to the part *AEB*.

For, suppose the part *AEB* to be turned around *AB*, until it shall fall on the part *ADB*. The curve *AEB* will then exactly coincide with the curve *ADB*, or else there would be some point in the curve *AEB* or *ADB*, unequally distant from the centre *C*, which is contrary to the definition of a circumference (Def. 24). Hence the two curves will be equal (Ax. 11).

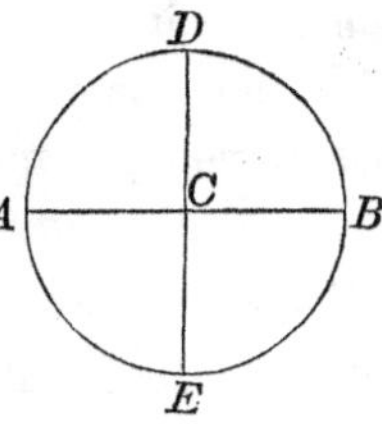

Corollary 1. If two lines, *AB*, *DE*, be drawn through the centre *C* perpendicular to each other, each will divide the circumference into two equal parts; and the entire circumference will be divided into the equal quadrants *DB*, *DA*, *AE*, and *EB*.

Cor. 2. Hence, a right angle, as *DCB*, is measured by one quadrant, or 90 degrees; two right angles by a semicircumference, or 180 degrees; and four right angles by the whole circumference, or 360 degrees

THEOREM II.

If one straight line meet another straight line, the sum of the two adjacent angles will be equal to two right angles.

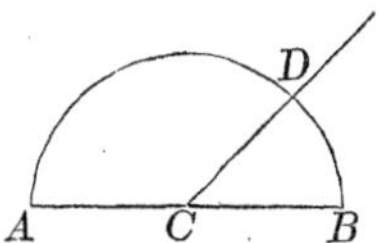

Let the straight line CD meet the straight line AB, at the point C; then will the angle DCB plus the angle DCA be equal to two right angles.

About the centre C, with any radius as CB, suppose a semicircumference to be described. Then, the angle DCB will be measured by the arc BD, and the angle DCA by the arc AD. But the sum of the two arcs is equal to a semicircumference: hence, the sum of the two angles is equal to two right angles (Th. i, Cor. 2).

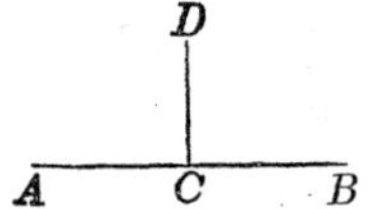

Cor. 1. If one of the angles, as DCB, is a right angle, the other angle, DCA will also be a right angle.

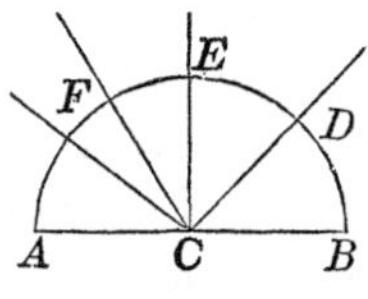

Cor. 2. Hence, all the angles which can be formed at any point C, by any number of lines, CD, CE, CF, &c., drawn on the same side of AB, will be equal to two right angles: for, they will be measured by the semicircumference $AFEDB$.

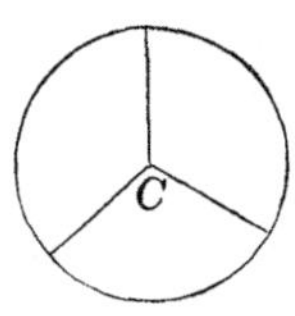

Cor. 3. Hence also, all the angles which can be formed round any point, as C, will be equal to four right angles. For, the sum of all the arcs which measure them, will be equal to the entire circumference, which is the measure of four right angles (Th. i, Cor. 2).

THEOREM III.

If two straight lines intersect each other, the opposite or vertical angles which they form, are equal.

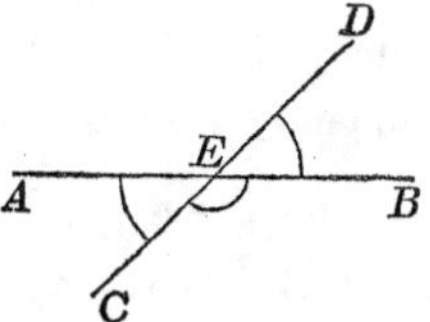

Let the two straight lines AB and CD intersect each other at the point E: then will the opposite angle AEC be equal to DEB, and $AED=CEB$.

For, since the line AE meets the line CD, the angle $AEC+AED=$ to two right angles. But since the line DE meets the line AB, we have $DEB+AED=$ two right angles. Taking away from these equals the common angle AED, and there will remain the angle AEC equal to the angle DEB (Ax. 3).

In the same manner we may prove that the angle AED is equal to the angle CEB.

THEOREM IV.

If two triangles have two sides and the included angle of the one, equal to two sides and the included angle of the other, each to each, the two triangles will be equal.

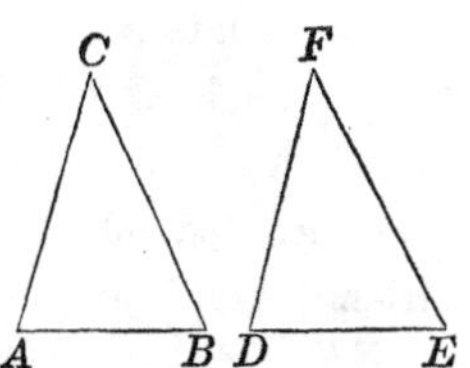

Let the triangles ABC and DEF have the side AC equal to DF, CB to FE, and the angle C equal to the angle F: then will the triangle ACB be equal to the triangle DEF.

For, suppose the side AC, of the triangle ACB, to be placed on DF, so that the extremity C shall fall on the extremity F: then, since the sides are equal A will fall on D.

But since the angle C is equal to the angle F, the line CB

will fall on FE; and since CB is equal to FE, the extremity B will fall on E; and consequently the side AB will fall on the side DE (Def. 3). Hence, the two triangles will fill the same space, and consequently are equal (Ax. 11.).

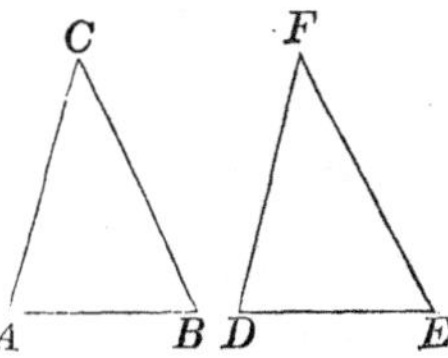

Scholium. Two triangles are said to be equal, when being applied to each other, they will exactly coincide (Ax. 11). Hence, *equal* triangles have their like parts equal, each to each, since those parts coincide with each other. The converse of the proposition is also true, namely, that *two triangles which have all the parts of the one equal to the corresponding parts of the other, each to each, are equal:* for if applied to each other, the equal parts will coincide.

THEOREM V.

If two triangles have two angles and the included side of the one, equal to two angles and the included side of the other, each to each, the two triangles will be equal.

Let the two triangles ABC and DEF have the angle A equal to the angle D, the angle B equal to the angle E, and the included side AB equal to the included side DE: then will the triangle ABC be equal to the triangle DEF.

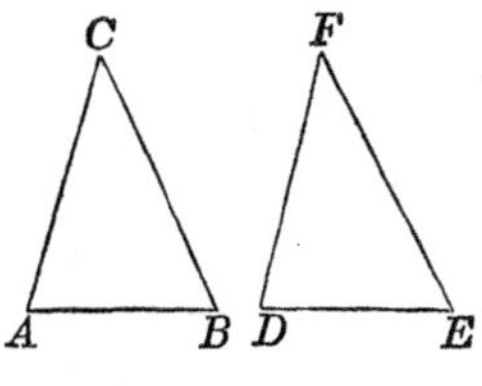

For, let the side AB be placed on the side DE, the extremity A on the extremity D; and since the sides are equal, the point B will fall on the point E.

Then, since the angle A is equal to the angle D, the side

AC will take the direction DF: and since the angle B is equal to the angle E, the side BC will fall on the side EF: hence, the point C will be found at the same time on DF and EF, and therefore will fall at the intersection F: consequently, all the parts of the triangle ABC will coincide with the parts of the triangle DEF, and therefore, the two triangles are equal.

THEOREM VI.

In an isosceles triangle the angles opposite the equal sides are equal to each other.

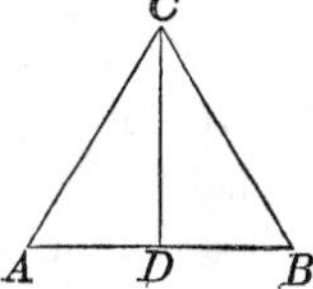

Let ABC be an isosceles triangle, having the side AC equal to the side CB: then will the angle A be equal to the angle B.

For, suppose the line CD to be drawn dividing the angle C into two equal parts.

Then, the two triangles ACD and DCB, have two sides and the included angle of the one equal to two sides and the included angle of the other, each to each: that is, the side AC equal to BC, the side CD common, and the included angle ACD equal to the included angle DCB: hence the two triangles are equal (Th. iv); and hence the angle A is equal to the angle B.

Cor. 1. Hence, the line which bisects the vertical angle of an isosceles triangle, bisects the base. It is also perpendicular to the base, since the angle CDA is equal to the angle CDB.

Cor. 2. Hence, also, every equilateral triangle, must also be equiangular: that is, have all its angles equal, each to each.

THEOREM VII.

Conversely.—*If a triangle has two of its angles equal, the sides opposite those angles will also be equal.*

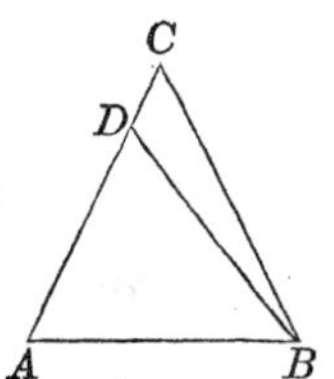

In the triangle ABC, let the angle A be equal to the angle B: then will the side BC be equal to the side AC.

For, if the two sides are not equal, one of them must be greater than the other. Suppose AC to be the greater side. Then take a part AD equal to BC.

Now, in the two triangles ADB and ABC, we have the side $AD=BC$, by hypothesis, the side AB common, and the angle A equal to the angle B: hence the two triangles have two sides and the included angle of the one equal to two sides and the included angle of the other, each to each: hence, the two triangles are equal (Th. iv), that is, a part ADB is equal to the whole ABC, which is impossible (Ax. 8): consequently, the side AC cannot be greater than the side CB, and hence, the triangle is isosceles.

Scholium 1. The method of reasoning pursued in the last theorem, is called the "reductio ad absurdum," or a proof that leads to a known absurdity.

Let us analyze this method of reasoning. We wished to prove that the two sides AC, CB were equal. We supposed them unequal, and AC the greater—that was an hypothesis (See Def. 12). We then reasoned on the hypothesis, and proved a part equal to the whole, which we know to be false (Ax. 8). Hence, we conclude that the hypothesis is untrue because after a correct chain of reasoning it leads to a resi which we know to be absurd.

Scholium 2. Generally,—If the demonstration is based on known principles, previously proved, or admitted in the axioms, the conclusion will always be true. But, if the demonstration is based on an hypothesis, (as in the last theorem, that AC was the greater side), and the conclusion is *contrary* to what has been previously proved, or admitted in the axioms, then, it follows, that the hypothesis cannot be true.

The former is called a ***positive***, and the latter a ***negative*** demonstration.

THEOREM VIII.

If two triangles have the three sides of the one equal to the three sides of the other, each to each, the three angles will also be equal, each to each.

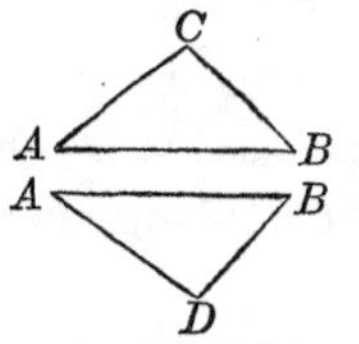

Let the two triangles ABC, ABD, have the side AB equal to the side AB, the side AC equal to AD, and the side CB equal to DB: then will the corresponding angles also be equal, viz: the angle A will be equal to the angle A, the angle B to the angle B, and the angle C to the angle D.

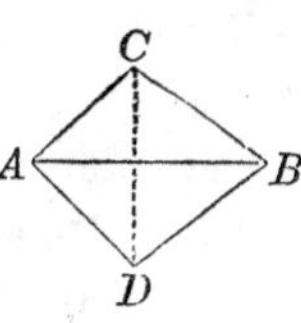

For, suppose the triangles to be joined by their longest equal sides AB, and the line CD to be drawn.

Then, since the side AC is equal to AD, by hypothesis, the triangle ADC will be isosceles; and therefore, the angle ACD will be equal to the angle ADC (Th. vi). In like manner in the triangle CBD, the side CB is equal to DB: hence, the angle BCD is equal to the angle BDC.

Now, by the addition of equals, we have

$$ACD+BCD=ADC+BDC$$

that is, the angle $ACB=ADB$.

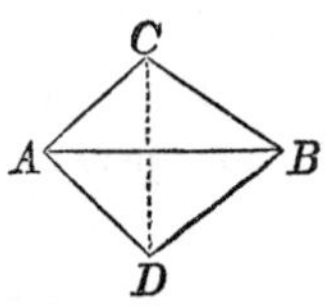

Now, the two triangles ACB and ADB have two sides and the included angle of the one equal to two sides and the included angle of the other, each to each: hence, the remaining angles will be equal (Th. iv): consequently, the angle CAB is equal to BAD, and the angle CBA to the angle ABD.

Sch. The angles of the two triangles which are equal to each other, are those which lie opposite the equal sides.

THEOREM IX.

If one side of a triangle is produced, the outward angle is greater than either of the inward opposite angles.

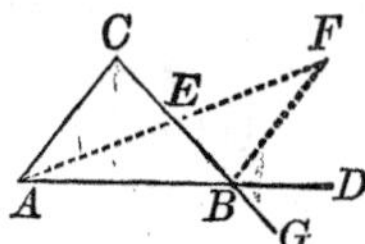

Let ABC be a triangle, having the side AB produced to D: then will the outward angle CBD be greater than either of the inward opposite angles A or C.

For, suppose the side CB to be bisected at the point E. Draw AE, and produce it until EF is equal to AE, and then draw BF.

Now, since the two triangles AEC and BEF have $AE=EF$ and $EC=EB$, and the included angle AEC equal to the included angle BEF (Th. iii), the two triangles will be equal in all respects (Th. iv): hence, the angle EBF will be equal to the angle C. But the angle CBD is greater than the angle CBF, consequently it is greater than the angle C.

In like manner, if CB be produced to G, and AB be bisected, it may be proved that the outward angle ABG, or its equal CBD (Th. iii), is greater than the angle A.

THEOREM X.

The sum of any two sides of a triangle is greater than the third side.

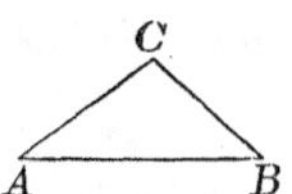

Let ABC be a triangle: then will the sum of two of its sides, as AC, CB, be greater than the third side AB.

For, the straight line AB is the shortest distance between the two points A and B (Def. 3): hence, $AC+CB$ is greater than AB.

THEOREM XI.

The greater side of every triangle is opposite the greater angle. and conversely, the greater angle is opposite the greater side.

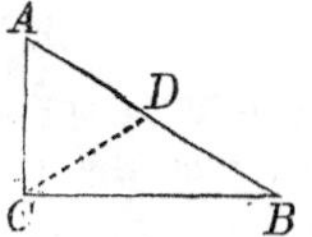

First. In the triangle CAB, let the angle C be greater than the angle B: then, will the side AB be greater than the side AC.

For, draw CD, making the angle BCD equal to the angle B. Then, the triangle CBD will be isosceles: hence, the side $CD=DB$ (Th. vi).

But, by the last theorem AC is less than $AD+CD$; that is, less than $AD+DB$, and consequently less than AB.

Secondly. Let us suppose the side AB to be greater than AC; then will the angle C be greater than the angle B.

For, if the angle C were equal to B, the triangle CAB would be isosceles, and the side AC would be equal to AB (Th. vi), which would be contrary to the hypothesis.

Again, if the angle C were less than B, then, by the first part of the theorem, the side AB would be less than AC, which is also contrary to the hypothesis Hence, since C

cannot be equal to B, nor less than B, it follows that it must be greater.

THEOREM XII.

If a straight line intersect two parallel lines, the alternate angles will be equal.

If two parallel straight lines, AB CD, are intersected by a third line GH, the angles AEF and EFD are called *alternate* angles. It is required to prove that these angles are equal.

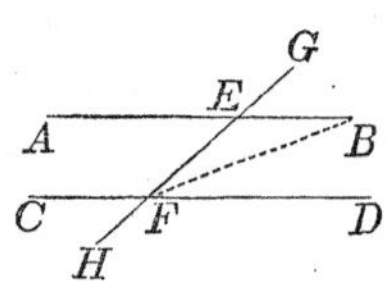

If they are unequal one of them must be greater than the other. Suppose EFD to be the greater angle.

Now conceive FB to be drawn, making the angle EFB equal to the angle AEF, and meeting AB in B.

Then, in the triangle FEB the outward angle FEA is greater than either of the inward angles B or EFB (Th. ix.); and therefore, EFB can never be equal to AEF so long as FB meets EB.

But since we have supposed EFD to be greater than AEF, it follows that EFB could not be equal to AEF, if FB fell below FD. Therefore, if the angle EFB is equal to the angle AEF, FB cannot meet AB, nor fall below FD, and consequently must coincide with the parallel CD (Def. 30): and hence, the alternate angles AEF and EFD are equal.

Cor. If a line be perpendicular to one of two parallel lines, it will also be perpendicular to the other.

THEOREM XIII.

Conversely,—*If a line intersect two straight lines, making the alternate angles equal, those straight lines will be parallel.*

Let the line EF meet the lines AB, CD, making the angle AEF equal to the angle EFD: then will the lines AB and CD be parallel.

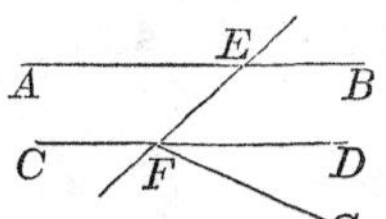

For, if they are not parallel, suppose through the point F the line FG to be drawn parallel to AB.

Then, because of the parallels AB, FG, the alternate angles, AEF and EFG will be equal (Th. xii). But, by hypothesis, the angle AEF is equal to EFD: hence, the angle EFD is equal to the angle EFG (Ax. 1); that is, a part is equal to the whole, which is absurd (Ax. 8): therefore no line but CD can be parallel to AB.

Cor. If two lines are perpendicular to the same line, they will be parallel to each other.

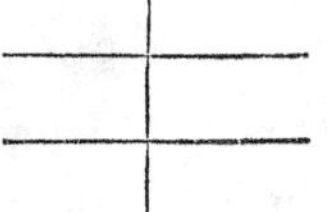

THEOREM XIV.

If a line cut two parallel lines, the outward angle is equal to the inward opposite angle on the same side; and the two inward angles, on the same side, are equal to two right angles.

Let the line EF cut the two parallels AB, CD: then will the outward angle EGB be equal to the inward opposite angle EHD; and the two inward angles, BGH and GHD, will be equal to two right angles.

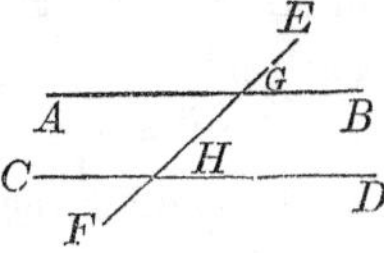

First. Since the lines AB, CD, are parallel, the angle AGH is equal to the alternate angle GHD (Th. xii); but the angle AGH is equal to the opposite angle EGB: hence, the angle EGB is equal to the angle EHD (Ax. 1).

Secondly. Since the two adjacent angles EGB and BGH are equal to two right angles (Th. ii); and since the angle EGB has been proved equal to EHD, it follows that the sum of BGH plus GHD, is also equal to two right angles.

Cor. 1. Conversely, if one straight line meets two other straight lines, making the angles on the same side equal to each other, those lines will be parallel.

Cor. 2. If a line intersect two other lines, making the sum of the two inward angles equal to two right angles, those two lines will be parallel.

Cor. 3. If a line intersect two other lines, making the sum of the two inward angles less than two right angles, those lines will not be parallel, but will meet if sufficiently produced.

THEOREM XV.

All straight lines which are parallel to the same line, are parallel to each other.

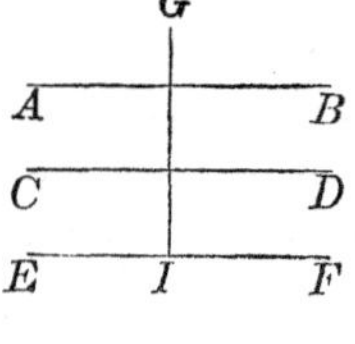

Let the lines AB and CD be each parallel to EF: then will they be parallel to each other.

For, let the line GI be drawn perpendicular to EF: then will it also be perpendicular to the parallels AB, CD (Th. xii Cor.).

Then, since the lines AB and CD are perpendicular to the line GI, they will be parallel to each other (Th. xiii. Cor).

THEOREM XVI.

If one side of a triangle be produced, the outward angle will be equal to the sum of the inward opposite angles.

In the triangle ABC, let the side AB be produced to D: then will the outward angle CBD be equal to the sum of the inward opposite angles A and C.

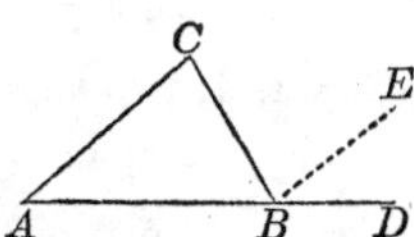

For, conceive the line BE to be drawn parallel to the side AC. Then, since BC meets the two parallels AC, BE, the alternate angles ACB and CBE will be equal (Th. xii).

And since the line AD cuts the two parallels BE and AC, the angles EBD and CAB are equal to each other (Th. xiv). Therefore, the inward angles C and A, of the triangle ABC, are equal to the angles CBE and EBD; and consequently, the sum of the two angles, A and C, is equal to the outward angle CBD (Ax. 1).

THEOREM XVII.

In any triangle the sum of the three angles is equal to two right angles.

Let ABC be any triangle: then will the sum of the three angles

$$A+B+C=\text{two right angles.}$$

For, let the side AB be produced to D. Then, the outward angle

$$CBD=A+C \text{ (Th. xvi).}$$

To each of these equals add the angle CBA, and we shall have

$$CBD+CBA=A+C+B.$$

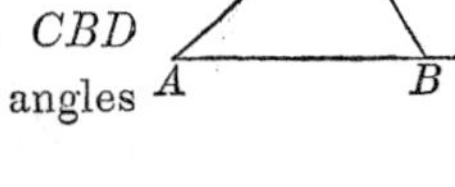

But the sum of the two angles CBD and CBA, is equal to two right angles (Th. ii): hence

$$A+B+C=\text{two right angles (Ax. 1).}$$

Cor. 1. If two angles of one triangle be equal to two angles of another triangle, the third angles will also be equal (Ax. 3).

Cor. 2. If one angle of one triangle be equal to one angle of another triangle, the sum of the two remaining angles in each triangle, will also be equal (Ax. 3).

Cor. 3. If one angle of a triangle be a right angle, the sum of the other two angles will be equal to a right angle; and each angle singly, will be acute.

Cor. 4. No triangle can have more than one right angle, nor more than one obtuse angle; otherwise, the sum of the three angles would exceed two right angles: hence, at least two angles of every triangle must be acute.

THEOREM XVIII.

I. *A perpendicular is the shortest line that can be drawn from a given point to a given line.*

II. *If any number of lines be drawn from the same point, those which are nearest the perpendicular are less than those which are more remote.*

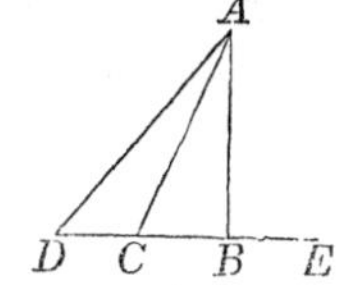

Let A be a given point, and DE a straight line. Suppose AB to be drawn perpendicular to DE, and suppose the oblique lines AC and AD also to be drawn: Then, AB will be shorter than either of the oblique lines, and AC will be less than AD.

First. Since the angle B, in the triangle ACB, is a right angle, the angle C will be acute (Th. xvii. Cor. 3): and since the less side of every triangle is opposite the less angle (Th. xi), the side AB will be less than AC.

Secondly. Since the angle ACB is acute, the adjacent angle ACD will be obtuse (Th. ii): consequently, the angle D is acute (Th. xvii. Cor. 3), and therefore less than the angle ACD. And since the less side of every triangle is opposite the less angle, it follows that AC is less than AD.

Cor. A perpendicular is the shortest distance from a point to a line.

THEOREM XIX.

If two right angled triangles have the hypothenuse and a side of the one equal to the hypothenuse and a side of the other, the remaining parts will also be equal, each to each.

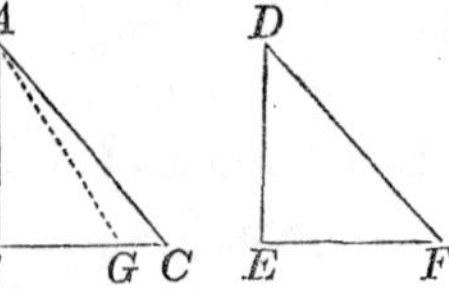

Let the two right angled triangles ABC and DEF, have the hypothenuse AC equal to DF, and the side AB equal to DE: then will the remaining parts be equal, each to each.

For, if the side BC is equal to EF, the corresponding angles of the two triangles will be equal (Th. viii). If the sides are unequal, suppose BC to be the greater, and take a part BG, equal to EF, and draw AG.

Then, in the two triangles ABG and DEF, the angle B is equal to the angle E, the side AB to the side DE, and the side BG to the side EF: hence, the two triangles are equal in all respects (Th. iv), and consequently, the side AG is equal to

DF. But DF is equal to AC, by hypothesis; therefore, AG is equal to AC (Ax 1). But this is impossible (Th. xviii); hence, the sides BC and EF cannot be unequal; consequently, the triangles are equal (Th. viii).

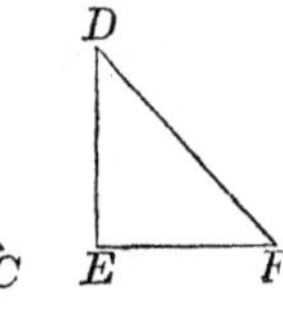

THEOREM XX.

The sum of the four angles of every quadrilateral is equal to four right angles.

Let $ACBD$ be a quadrilateral: then will

$A+B+C+D=$four right angles.

Let the diagonal DC be drawn dividing the quadrilateral AB, into two triangles, BDC, ADC.

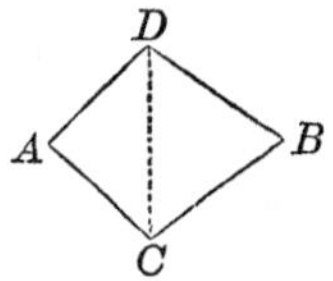

Then, because the sum of the three angles of each triangle is equal to two right angles (Th. xvii), it follows that the sum of the angles of both triangles is equal to four right angles. But the sum of the angles of both triangles, make up the angles of the quadrilateral. Hence, the sum of the four angles of the quadrilateral is equal to four right angles.

Cor. 1. If then three of the angles be right angles, the fourth angle will also be a right angle.

Cor. 2. If the sum of two of the four angles be equal to two right angles, the sum of the remaining two will also be equal to two right angles.

THEOREM XXI.

The sum of all the interior angles of any polygon is equal to twice as many right angles, wanting four, as the figure has sides.

Let $ABCDE$ be any polygon: then will the sum of its inward angles

$$A+B+C+D+E$$

be equal to twice as many right angles, wanting four, as the figure has sides.

For, from any point P, within the polygon, draw the lines PA, PB, PC, PD, PE, to each of the angles, dividing the polygon into as many triangles as the figure has sides.

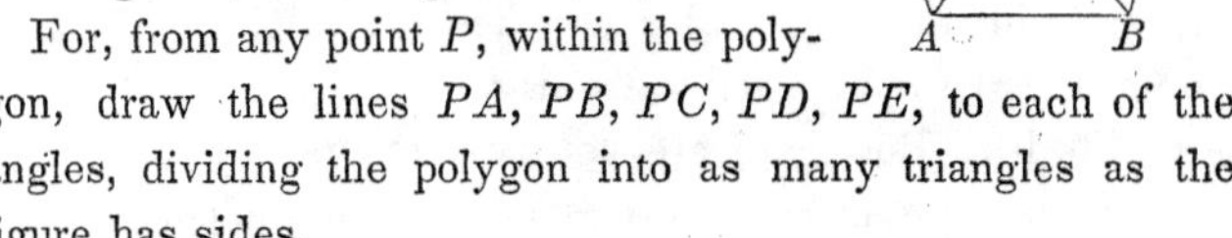

Now, the sum of the three angles of each of these triangles is equal to two right angles (Th. xvii): hence, the sum of the angles of all the triangles is equal to twice as many right angles as the figure has sides.

But the sum of all the angles about the point P is equal to four right angles (Th. ii. Cor. 3); and since this sum makes no part of the inward angles of the polygon, it must be subtracted from the sum of all the angles of the triangles, before found. Hence, *the sum of the interior angles of the polygon is equal to twice as many right angles, wanting four, as the figure has sides.*

Sch. This proposition is not applicable to polygons which have *re-entrant* angles.

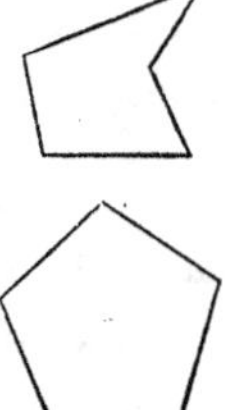

The reasoning is limited to polygons with salient angles, which may properly be named *convex polygons.*

THEOREM XXII.

If every side of a polygon be produced out, the sum of all the outward angles thereby formed, will be equal to four right angles.

Let A, B, C, D, and E, be the outward angles of a polygon formed by producing all the sides. Then will

$$A+B+C+D+E=\text{four right angles.}$$

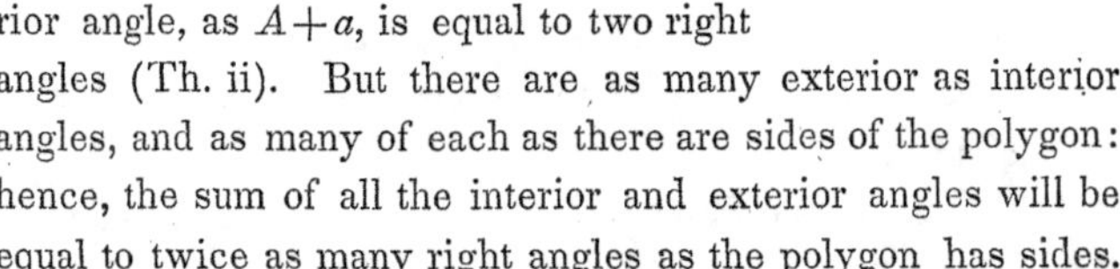

For, each interior angle, plus its exterior angle, as $A+a$, is equal to two right angles (Th. ii). But there are as many exterior as interior angles, and as many of each as there are sides of the polygon: hence, the sum of all the interior and exterior angles will be equal to twice as many right angles as the polygon has sides.

But the sum of all the interior angles together with four right angles, is equal to twice as many right angles as the polygon has sides (Th. xxi): that is, equal to the sum of all the inward and outward angles taken together.

From each of these equal sums take away the inward angles, and there will remain, the outward angles equal to four right angles (Ax. 3).

THEOREM XXIII.

The opposite sides and angles of every parallelogram are equal each to each: and a diagonal divides the parallelogram into two equal triangles.

Let $ABCD$ be any parallelogram, and DB a diagonal: then will the opposite sides and angles be equal to each other, each to each, and the diagonal DB will divide the parallelogram into two equal triangles.

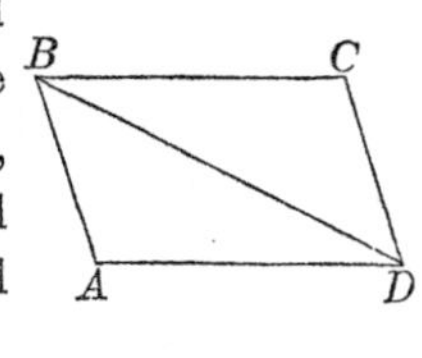

For, since the figure is a parallelogram, the sides AB, DC are parallel, as also the sides AD, BC. Now, since the parallels are cut by the diagonal DB, the alternate angles will be equal (Th. xii): that is the angle

$$ADB=DBC \quad \text{and} \quad BDC=ABD.$$

Hence, the two triangles ADB, BDC, having two angles in the one equal to two angles in the other, will have their third angles equal (Th. xvii. Cor. 1), viz. the angle A equal to the angle C, and these are two of the opposite angles of the parallelogram.

Also, if to the equal angles ADB, DBC, we add the equals BDC, ABD, the sums will be equal (Ax. 2): viz. the whole angle ADC to the whole angle ABC, and these are the other two opposite angles of the parallelogram.

Again, since the two triangles ADB, DBC, have the side DB common, and the two adjacent angles in the one equal to the two adjacent angles in the other, each to each, the two triangles will be equal (Th. v): hence, the diagonal divides the parallelogram into two equal triangles.

Cor. 1. If one angle of a parallelogram be a right angle, each of the angles will also be a right angle, and the parallelogram will be a rectangle.

Cor. 2. Hence, also, the sum of either two adjacent angles of a parallelogram, will be equal to two right angles.

THEOREM XXIV.

If the opposite sides of a quadrilateral, are equal, each to each, the equal sides will be parallel, and the figure will be a parallelogram

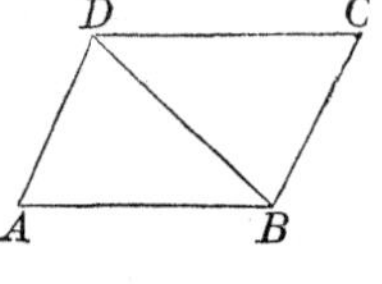

Let $ABCD$ be a quadrilateral, having its opposite sides respectively equal, viz.

$AB = CD$ and $AD = BC$

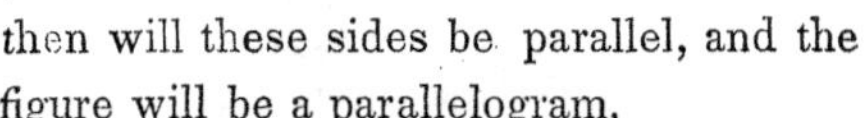

then will these sides be parallel, and the figure will be a parallelogram.

For, draw the diagonal BD. Then, the two triangles ABD, BDC, have all the sides of the one equal to all the sides of the other, each to each: therefore, the two triangles are equal (Th. viii); hence, the angle ADB, opposite the side AB, is equal to the angle DBC opposite the side DC; therefore, the sides AD, BC, are parallel (Th. xiii). For a like reason DC is parallel to AB, and the figure $ABCD$ is a parallelogram.

THEOREM XXV.

If two opposite sides of a quadrilateral are equal and parallel, the remaining sides will also be equal and parallel, and the figure will be a parallelogram.

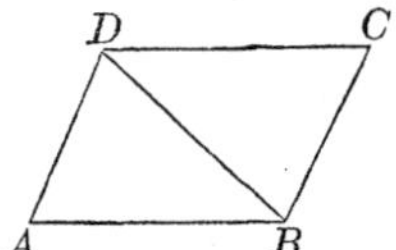

Let $ABCD$ be a quadrilateral, having the sides AB, CD, equal and parallel: then will the figure be a parallelogram.

For, draw the diagonal DB, dividing the quadrilateral into two triangles. Then, since AB is parallel to DC, the alternate angles, ABD and BDC are equal (Th. xii): moreover, the side BD is common; hence the two triangles have two sides and the included angle of the one, equal to two sides and the included angle of the other: the triangles are therefore equal, and consequently, AD is equal to BC, and the angle ADB to the angle DBC; and consequently, AD is also parallel to BC (Th. xiii) Therefore, the figure $ABCD$ is a parallelogram.

THEOREM XXVI.

The two diagonals of a parallelogram divide each other into equal parts, or mutually bisect each other

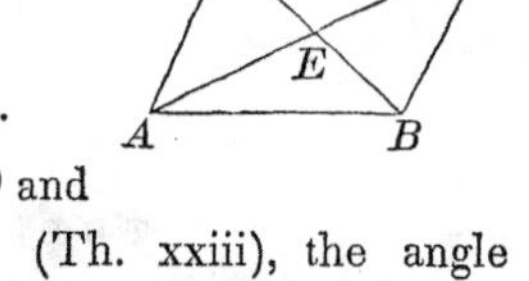

Let $ABCD$ be a parallelogram, and AC, BD its two diagonals intersecting at E. Then will

$AE=EC$ and $BE=ED$.

Comparing the two triangles AED and BEC, we find the side $AD=BC$ (Th. xxiii), the angle $ADE=EBC$ and $EAD=ECB$: hence, the two triangles are equal (Th. v): therefore, AE, the side opposite ADE, is equal to EC, the side opposite EBC; and ED is equal to EB.

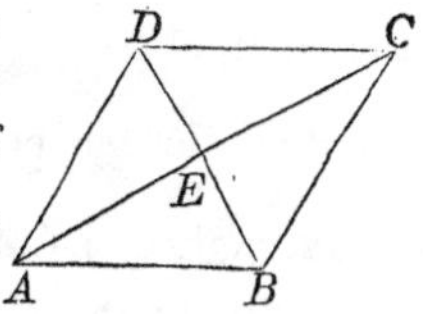

Sch. In the case of a rhombus (Def. 48), the sides AB, BC being equal, the triangles AEB and BEC have all the sides of the one equal to the corresponding sides of the other, and are therefore equal. Whence it follows that the angles AEB and BEC are equal. Therefore, the diagonals of a rhombus bisect each other at right angles.

GEOMETRY.

BOOK II,

OF THE CIRCLE.

DEFINITIONS.

1. The circumference of a circle is a curve line, all the points of which are equally distant from a certain point within called the centre.

2. The circle is the space bounded by this curve line.

3. Every straight line, CA, CD, CE, drawn from the centre to the circumference, is called a *radius* or *semidiameter*. Every line which, like AB, passes through the centre and terminates in the circumference, is called a *diameter*.

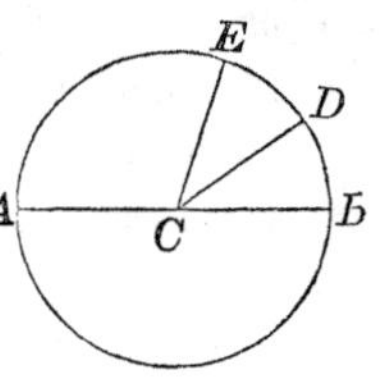

4. Any portion of the circumference, as EFG, is called an *arc*.

5. A straight line, as EG, joining the extremities of an arc, is called a *chord*.

6. A *segment* is the surface or portion of a circle included between an arc and its chord. Thus, EFG is a segment.

7. A *sector* is the part of the circle included between an arc and the two radii drawn through its extremities. Thus, CAB is a sector.

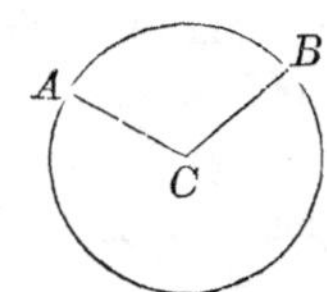

8. A straight line is said to be inscribed in a circle, when its extremities are in the circumference. Thus, the line AB is inscribed in a circle.

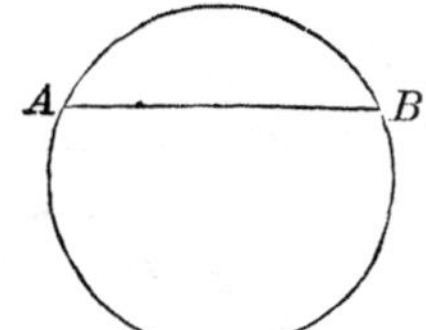

9. An inscribed angle is one which is formed by two chords that intersect each other in the circumference. Thus, BAC is an inscribed angle.

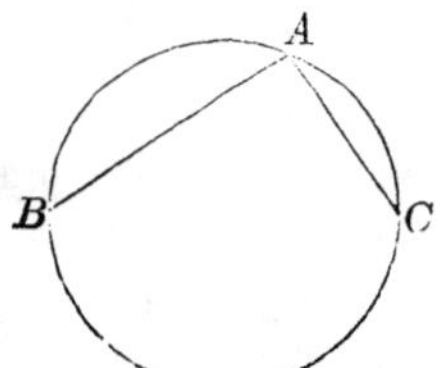

10. An inscribed triangle is one which has its three angular points in the circumference. Thus, ABC is an inscribed triangle.

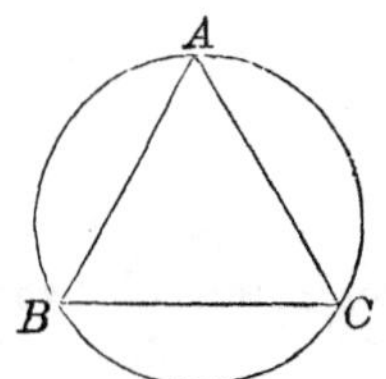

11. Any polygon is said to be inscribed in a circle when the vertices of all the angles are in the circumference. The circle is then said to circumscribe the polygon.

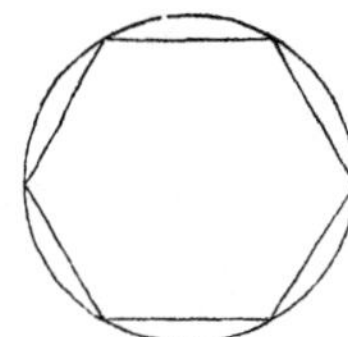

12. A *secant* is a line which meets the circumference in two points, and lies partly within and partly without the circle. Thus, AB is a secant.

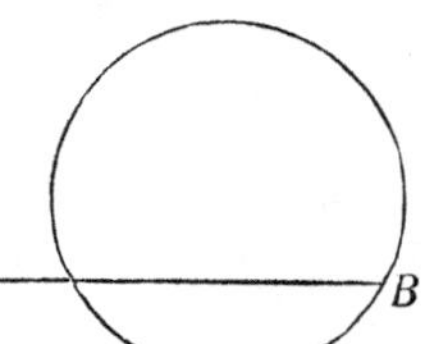

13. A *tangent* is a line which has but one point in common with the circumference. Thus, CMB is a tangent.

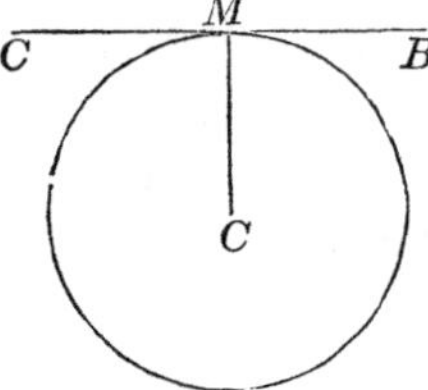

14. Two circles are said to touch each other internally, when one lies within the other, and their circumferences have but one point in common.

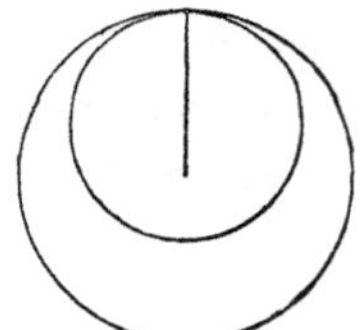

15. Two circles are said to touch each other externally, when one lies without the other, and their circumferences have but one point in common.

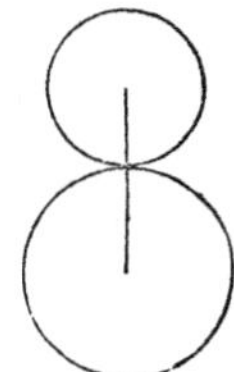

Of the Circle.

THEOREM I.

Every chord is less than a diameter.

Let AD be any chord. Draw the radii CA, CD to its extremities. We shall then have, AD less than $AC+CD$ (Book I. Th. x*). But $AC+CD$ is equal to the diameter AB: hence, the chord AD is less than the diameter.

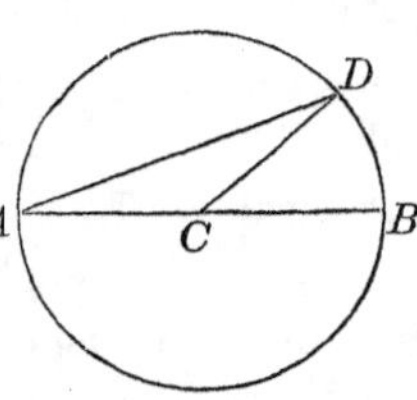

THEOREM II.

If from the centre of a circle a line be drawn to the middle of a chord,

I. *It will be perpendicular to the chord;*

II. *And it will bisect the arc of the chord.*

Let C be the centre of a circle, and AB any chord. Draw CD through D, the middle point of the chord, and produce it to E: then will CD be perpendicular to the chord, and the arc AE equal to EB.

First. Draw the two radii CA, CB. Then the two triangles ACD, DCB, have the three sides of the one equal to the three sides of the

*Note. When reference is made from one theorem to another, in the same Book, the number of the theorem referred to is alone given; but when the theorem referred to is found in a preceding Book, the number of the Book is also given.

other, each to each: viz. AC equal to CB, being radii, AD equal to DB, by hypothesis, and CD common: hence, the corresponding angles are equal (Book I. Th. viii): that is, the angle CDA equal to CDB, and the angle ACD equal to the angle DCB.

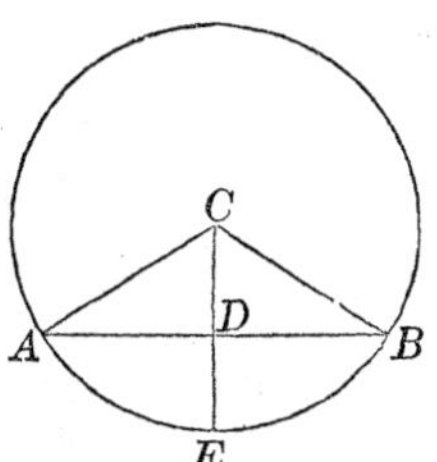

But, since the angle CDA is equal to the angle CDB, the radius CE is perpendicular to the chord AB (Bk. I. Def. 20).

Secondly. Since the angle ACE is equal to BCE, the arc AE will be equal to the arc EB, for equal angles must have equal measures (Bk. I. Def. 28).

Hence, the radius drawn through the middle point of a chord, is perpendicular to the chord, and bisects the arc of the chord.

Cor. Hence, a line which bisects a chord at right angles, bisects the arc of the chord, and passes through the centre of the circle. Also, a line drawn through the centre of the circle and perpendicular to the chord bisects it.

THEOREM III.

If more than two equal lines can be drawn from any point within a circle to the circumference, that point will be the centre.

Let D be any point within the circle ABC. Then, if the three lines DA, DB, and DC, drawn from the point D to the circumference, are equal, the point D will be the centre.

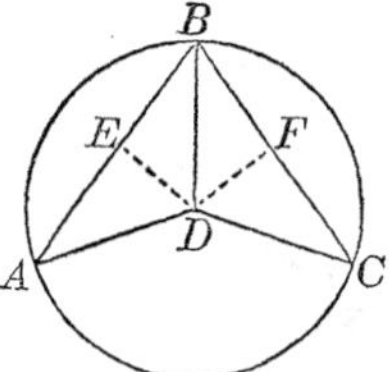

For, draw the chords AB, BC, bisect them at the points E and F, and join DE and DF.

Then, since the two triangles DAE and DEB have the side AE equal to EB, AD equal to DB, and DE common, they will be equal in all respects; and consequently, the angle DEA is equal to the angle DEB (Bk. I. Th. viii); and therefore, DE is perpendicular to AB (Bk. I. Def. 20). But, if DE bisects AB at right angles, it will pass through the centre of the circle (Th. ii. Cor).

In like manner, it may be shown that DF passes through the centre of the circle, and since the centre is found in the two lines ED, DF, it will be found at their common intersection D.

THEOREM IV.

Any chords which are equally distant from the centre of a circle, are equal.

Let AB and ED be two chords equally distant from the centre C: then will the two chords AB, ED be equal to each other.

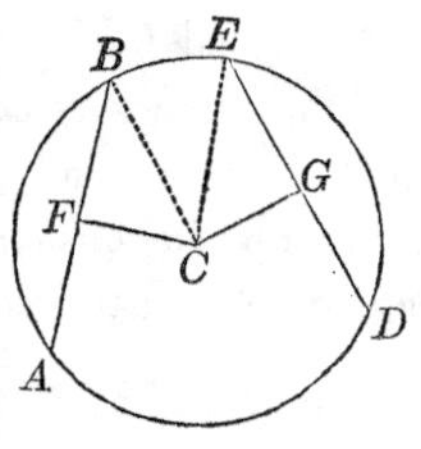

Draw CF perpendicular to AB, and CG perpendicular to ED, and since these perpendiculars measure the distances from the centre, they will be equal. Also draw CB and CE.

Then, the two right angled triangles CFB and CEG having the hypothenuse CB equal to the hypothenuse CE, and the side CF equal to CG, will have the third side BF equal to EG (Bk. I Th. xix). But, BF is the half of BA, and EG the half of DE (Th. ii. Cor); hence, BA is equal to DE Ax. 6).

THEOREM V.

A line which is perpendicular to a radius at its extremity, is tangent to the circle.

Let the line ABD be perpendicular to the radius CB at the extremity B: then will it be tangent to the circle at the point B.

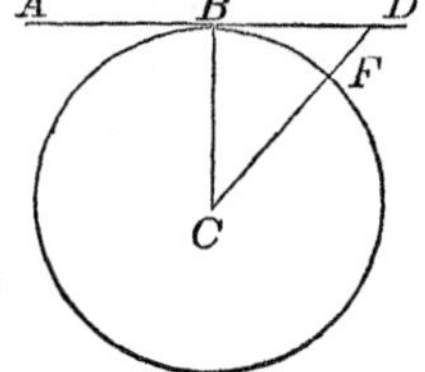

For, from any other point of the line, as D, draw DFC to the centre, cutting the circumference in F.

Then, because the angle B, of the triangle CDB, is a right angle, the angle at D is acute (Bk. I. Th. xvii. Cor. 3), and consequently less than the angle B. But the greater side of every triangle is opposite to the greater angle (Bk. I. Th. xi); therefore, the side CD is greater than CB, or its equal CF. Hence, the point D is without the circle, and the same may be shown for every other point of the line AD. Consequently, the line ABD has but one point in common with the circumference of the circle, and therefore is tangent to it at the point B (Def. 13).

Cor. Hence, if a line is tangent to a circle, and a radius be drawn through the point of contact, the radius will be perpendicular to the tangent.

THEOREM VI.

If the distance between the centres of two circles is equal to the sum of their radii, the two circles will touch each other externally.

Let C and D be the two centres, and suppose the distance between them to be equal to the sum of the radii, that is, to $CA+AD$.

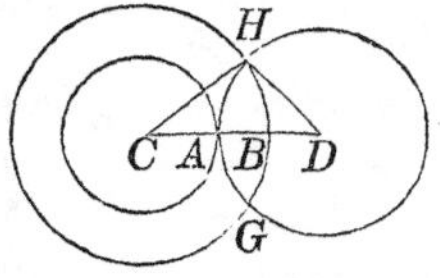

The circumferences of the circles will evidently have the point A common, and they will have no other. Because, if they had two points common, that, is if they cut each other in two points, G and H, the distance CD between their centres would be less than the sum of their radii CH, HD (Bk. I. Th. x); but this would be contrary to the supposition.

THEOREM VII.

If the distance between the centres of two circles is equal to the difference of their radii, the two circles will touch each other internally.

Let C and D be the centres of two circles at a distance from each other equal to $AD-AC=CD$.

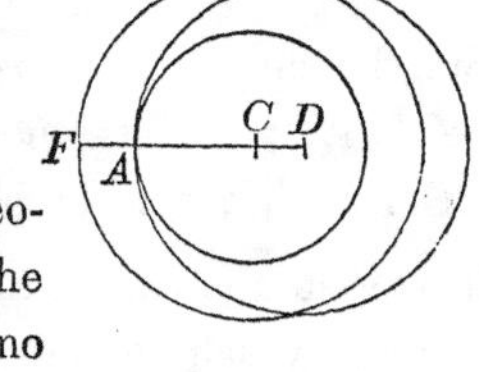

Now, it is evident, as in the last theorem, that the circumferences will have the point A common; and they can have no other. For, if they had two points common, the difference between the radii AD and FC would not be equal to CD, the distance between their centres: therefore, they cannot have two points in common when the difference of their radii is equal to the distance between their centres: hence, they are tangent to each other.

Sch. If two circles touch each other, either externally or internally, their centres and the point of contact will be in the same straight line

THEOREM VIII.

An angle at the circumference of a circle is measured by half the arc that subtends it.

Let BAD be an inscribed angle: then will it be measured by half the arc BED, which subtends it.

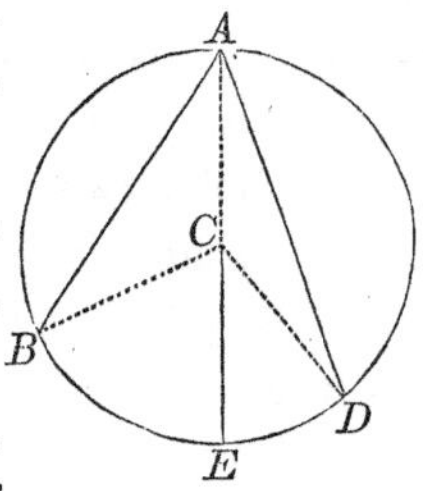

For, through the centre C draw the diameter ACE, and draw the radii BC, CD.

Then, in the triangle ABC, the exterior angle BCE is equal to the sum of the interior angles B and A (Bk. I. Th. xvi). But since the triangle BAC is isosceles, the angles A and B are equal (Bk. I. Th. vi); therefore, the exterior angle BCE is equal to double the angle BAC.

But, the angle BCE is measured by the arc BE, which subtends it; and consequently, the angle BAE, which is half of BCE, is measured by half the arc BE.

It may be shown, in like manner, that the angle EAD is measured by half the arc ED: and hence, by the addition of equals, it would follow that, the angle BAD is measured by half the arc BED, which subtends it.

Cor. 1. Hence, if an angle at the centre, and an angle at the circumference, both stand on the same arc, the angle at the centre will be double the angle at the circumference.

Cor. 2. If two angles at the circumference stand on equal arcs they will be equal to each other.

THEOREM IX.

All angles at the circumference, which stand upon the same arc, are equal to each other.

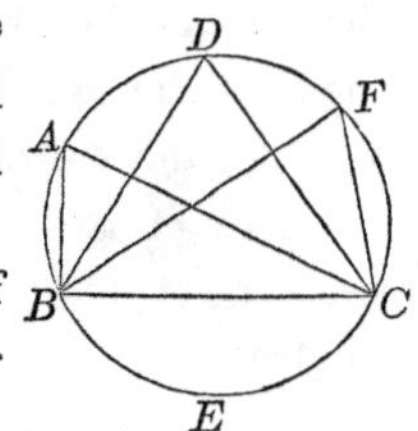

Let the angles BAC, BDC, BFC, have their vertices in the circumference, and stand on the same arc BEC: then will they be equal to each other.

For, each angle is measured by half the arc BEC (Th. viii); hence, the angles are all equal.

THEOREM X.

An angle in a semicircle, is a right angle.

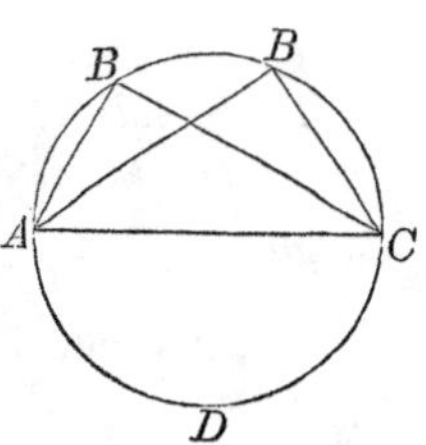

Let $ABBC$ be a semicircle: then will every angle, as B, B, inscribed in it, be a right angle.

For, each angle is measured by half the semicircumference ADC, that is, by a quadrant, which measures a right angle (Bk. I. Th. i. Cor. 2).

THEOREM XI.

If a quadrilateral be inscribed in a circle, the sum of either two of its opposite angles is equal to two right angles.

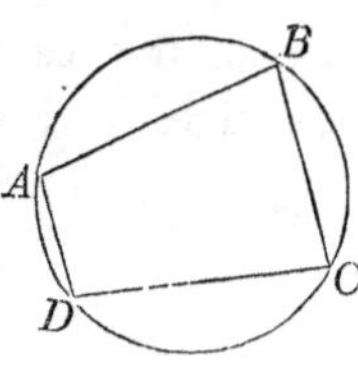

Let $ABCD$ be any quadrilateral inscribed in a circle; then will the sum of the two opposite angles, A and C, or B and D, be equal to two right angles.

For, the angle A is measured by half the arc DCB, which subtends it (Th. viii);

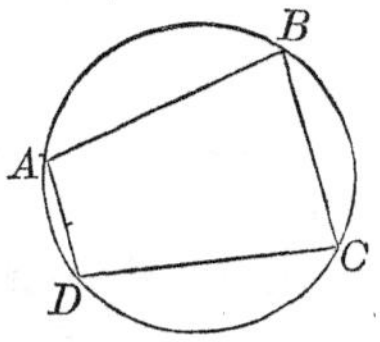

and the angle C is measured by half the arc DAB, which subtends it. Hence, the sum of the two angles, A and C, is measured by half the entire circumference. But half the entire circumference is the measure of two right angles; therefore, the sum of the opposite angles A and C is equal to two right angles.

In like manner, it may be shown, that the sum of the two angles B and D is equal to two right angles

THEOREM XII.

If the side of a quadrilateral, inscribed in a circle, be produced out, the exterior angle will be equal to the inward opposite angle.

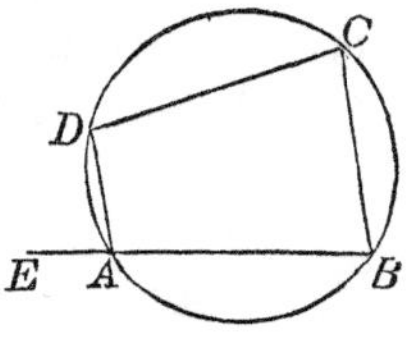

Let the side BA, of the quadrilateral $ABCD$ be produced to E, then will the outward angle DAE be equal to the inward opposite angle C.

For, the angle DAB plus the angle C, is equal to two right angles (Th. xi). But DAB plus DAE is also equal to two right angles (Bk. I. Th. ii). Taking from each the common angle DAB, and we shall have the angle DAE equal to the interior opposite angle C.

THEOREM XIII.

Two parallel chords intercept equal arcs.

Let the chords AB and CD be parallel: then will the arcs AC and BD be equal.

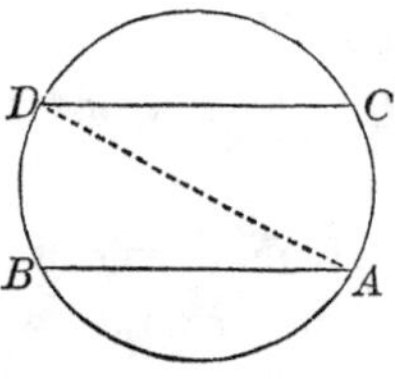

For, draw the line AD. Then, because the lines AB and CD are parallel, the alternate angles ADC and DAB will be equal (Bk. I. Th. xii). But the angle ADC is measured by half the arc AC, and the angle DAB by half the arc BD (Th. viii): hence, the two arcs AC and BD are themselves equal.

THEOREM XIV.

The angle formed by a tangent and a chord, is measured by half the arc of the chord.

Let BAE be tangent to the circle at the point A, and AC any chord.

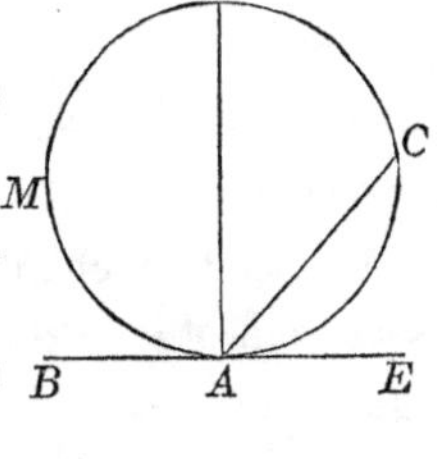

From A, the point of contact, draw the diameter AD.

Then, the angle BAD will be a right angle (Th. v. Cor), and therefore will be measured by half the semicircle AMD (Bk. I, Th. i. Cor. 2).

But the angle DAC being at the circumference, is measured by half the arc DC: hence, by the addition of equals, the two angles BAD and DAC, or the entire angle BAC will be measured by half the arc $AMDC$.

It may be shown, by taking the difference between the two angles DAE and DAC, that the angle CAE is measured by half the arc AC included between its sides.

5

THEOREM XV.

If a tangent and a chord are parallel to each other, they will intercept equal arcs.

Let the tangent ABC be parallel to the chord DF: then will the intercepted arcs DB, BF, be equal to each other.

For, draw the chord DB. Then, since AC and DF are parallel, the angle ABD will be equal to the angle BDF. But ABD being formed by a tangent and a chord, will be measured by half the arc DB; and BDF being an angle at the circumference will be measured by half the arc BF (Th. viii). But since the angles are equal, the arcs will be equal: hence DB is equal to BF.

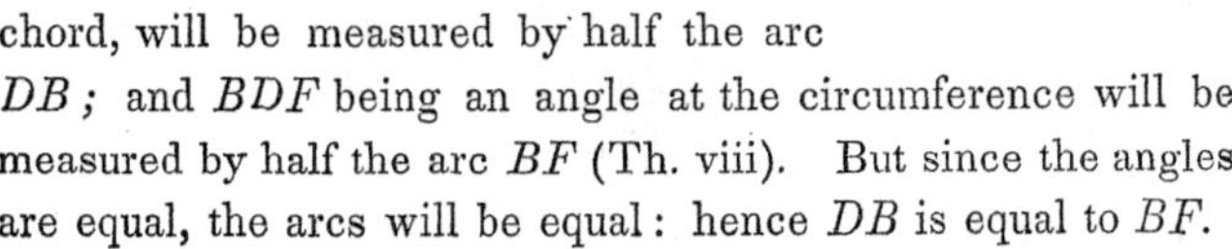

THEOREM XVI.

The angle formed within a circle by the intersection of two chords, is measured by half the sum of the intercepted arcs.

Let the two chords AB and CD intersect each other at the point E: then will the angle AEC, or its equal DEB, be measured by half the sum of the intercepted arcs AC, DB.

For, draw the chord AF parallel to CD. Then because of the parallels, the ngle DEB will be equal to the angle FAB (Bk I. Th. xiv), and the arc FD to the arc AC. But the angle FAB is measured by half the arc FDB, that is, by half the sum of the arcs FD, DB. Now, since FD is equal to AC, it follows that the angle DEB, or its equal AEC, will be measured by half the sum of the arcs DB and AC.

THEOREM XVII.

The angle formed without a circle by the intersection of two secants is measured by half the difference of the intercepted arcs.

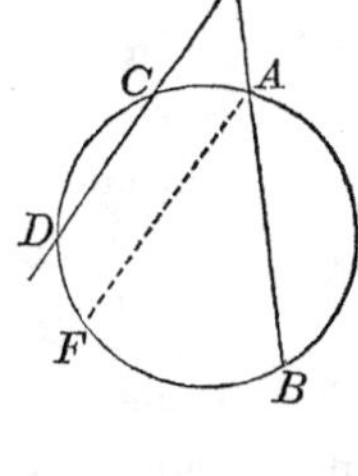

Let the two secants DE and EB intersect each other at E: then will the angle DEB be measured by half the intercepted arcs CA and DB.

Draw the chord AF parallel to ED. Then, because AF and ED are parallel, and EB cuts them, the angles FAB and and DEB are equal (Bk. I. Th. xiv).

But the angle FAB, at the circumference, is measured by half the arc FB (Th. viii), which is the difference of the arcs DFB and CA: hence, the equal angle E is also measured by half the difference of the intercepted arcs DFB and CA.

THEOREM XVIII.

An angle formed by two tangents is measured by half the difference of the intercepted arcs.

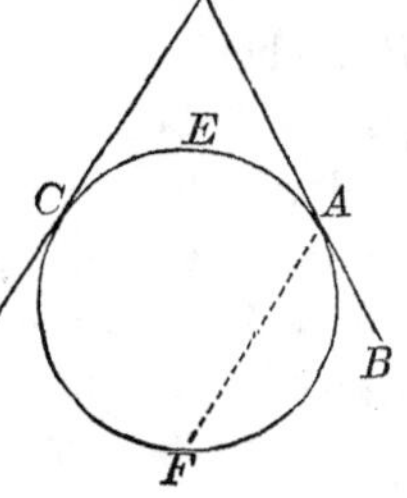

Let CD and DA be two tangents to the circle at the points C and A: then will the angle CDA be measured by half the difference of the intercepted arcs CEA and CFA.

For, draw the chord AF parallel to the tangent CD. Then, because the lines CD and AF are parallel, the angle BAF will be equal to the angle BDC (Bk. I. Th. xiv). But the angle BAF, formed by a tangent and a chord, is measured by half the arc AF, that is, by half the difference of CFA and CF.

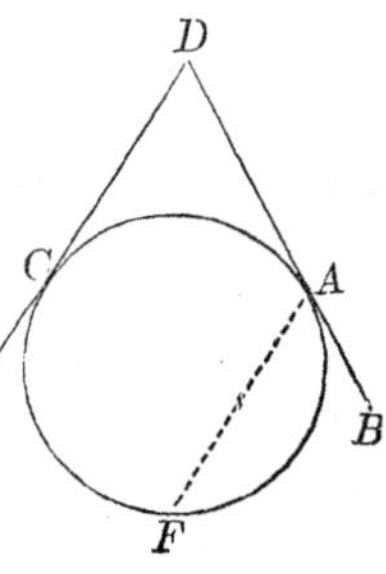

But since the tangent DC and the chord AF are parallel, the arc CF is equal to the arc CA: hence the angle BAF, or its equal BDC, which is measured by half the difference of CFA and CF, is also measured by half the difference of the intercepted arcs CFA and CA.

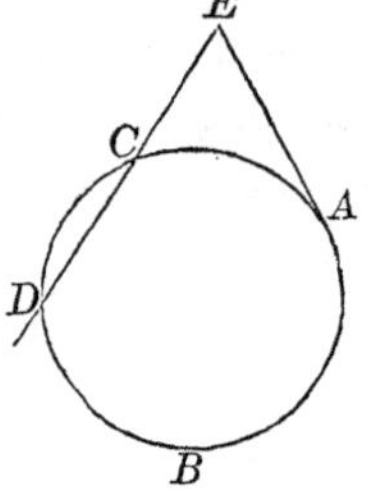

Cor. In like manner it may be proved that the angle E, formed by a tangent and secant, is measured by half the difference of the intercepted arcs AC and DBA.

THEOREM XIX.

The chord of an arc of sixty degrees is equal to the radius of the circle.

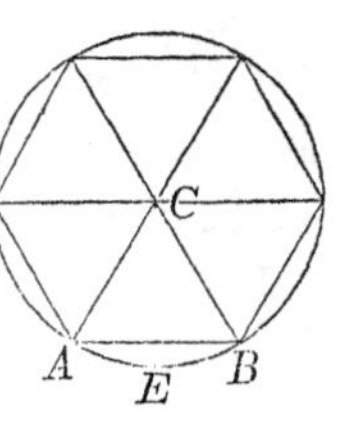

Let AEB be an arc of sixty degrees and AB its chord: then will AB be equal to the radius of the circle.

For, draw the radii CB and CA. Then, since the angle ACB is at the centre, it will be measured by the arc AEB: that is, it will be equal to sixty degrees (Bk. I. Def. 29).

Again, since the sum of the three angles of a triangle is equal to one hundred and eighty degrees (Bk. I. Th. xvii), it follows that the sum of the two angles A and B will be equal to one hundred and twenty degrees. But the triangle CAB is isosceles: hence, the angles at the base are equal (Bk. I. Th. vi): hence, each angle is equal to sixty degrees, and consequently, the side AB is equal to AC or CB (Bk. I. Th. vi).

PROBLEMS

RELATING TO THE FIRST AND SECOND BOOKS.

THE Problems of Geometry explain the methods of constructing or describing the geometrical figures.

For these constructions, a straight ruler and the common compasses or dividers, are all the instruments that are absolutely necessary.

DIVIDERS OR COMPASSES.

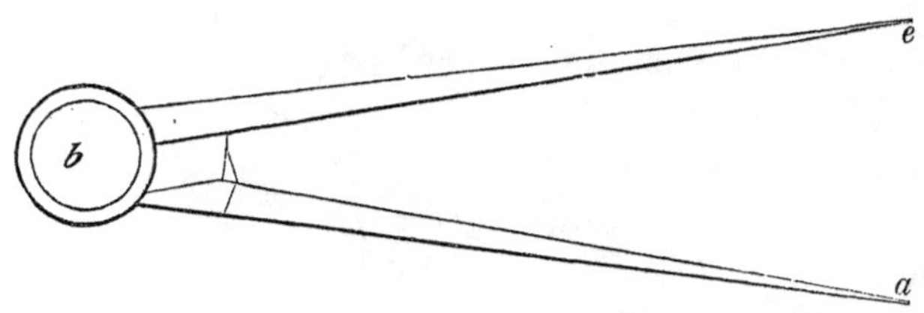

The dividers consist of the two legs *ba*, *be*, which turn easily about a common joint at *b*. The legs of the dividers are extended or brought together by placing the forefinger on the joint at *b*, and pressing the thumb and fingers against the legs.

PROBLEM I.

On any line, as CD, to lay off a distance equal to AB

Take up the dividers with the thumb and second finger, and place the forefinger on the joint at *b*.

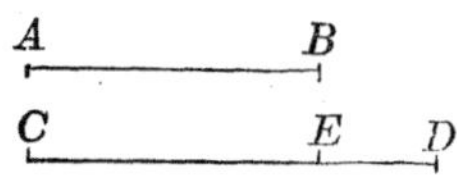

Then, set one foot of the dividers at *A*, and extend the legs with the thumb and fingers, until the other foot reaches *B*.

Then, raise the dividers, place one foot at *C*, and mark with the other the distance *CE:* and this distance will evidently be equal to *AB*.

PROBLEM II.

To describe from a given centre the circumference of a circle having a given radius.

Let *C* be the given centre, and *CB* the given radius.

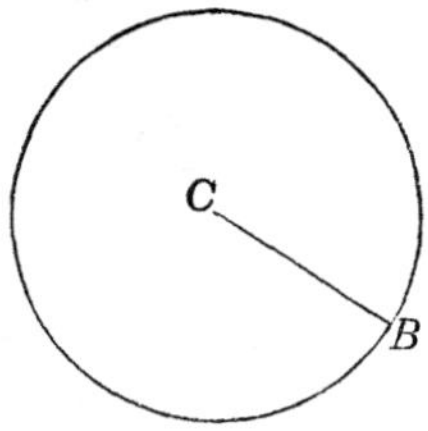

Place one foot of the dividers at *C*, and extend the other leg until it reaches to B. Then, turn the dividers around the leg at *C*, and the other leg will describe the required circumference.

OF THE RULER.

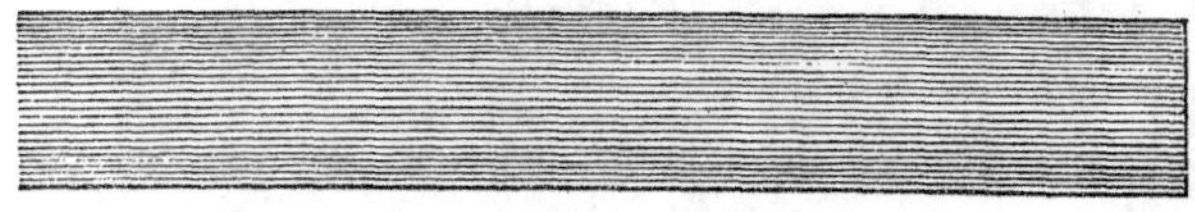

A ruler of a convenient size, is about twenty inches ir length, two inches wide, and one fifth of an inch in thickness It should be made of a hard material, and perfectly straigh and smooth.

PROBLEM III.

To draw a straight line through two given points A and B

Place one edge of the ruler on A and slide the ruler around until the same edge falls on B. Then, with a pen, or pencil, draw the line AB.

PROBLEM IV.

To bisect a given line: that is, to divide it into two equal parts

Let AB be the given line to be divided. With A as a centre, and radius greater than half of AB, describe an arc IFE. Then, with B as a centre, and an equal radius BI, describe the arc IHE. Join the points I and E by the line IE: the point D, where it intersects AB, will be the middle point of the line AB.

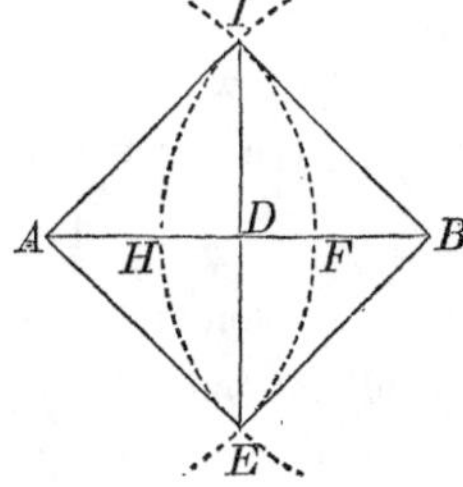

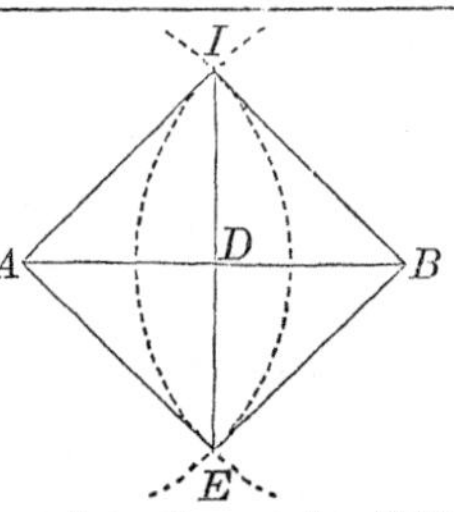

For, draw the radii AI, AE, BI, and BE. Then, since these radii are equal, the triangles AIE nd BIE have all the sides of the one equal to the corresponding sides of the other; hence, their corresponding angles are equal (Bk. I. Th. viii); that is, the angle AIE is equal to the angle BIE. Therefore, the two triangles AID and BID, have the side $AI=IB$, the angle $AID=BID$, and ID common: hence, they are equal (Bk. I. Th. iv), and AD is equal to DB

PROBLEM V.

To bisect a given angle or a given arc.

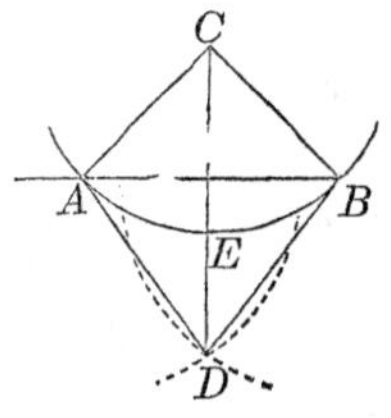

Let ACB be the given angle, and AEB the given arc.

From the points A and B as centres, describe with the same radius two arcs cutting each other in D. Through D and the centre C, draw CED, and it will divide the angle ACB into two equal parts, and also bisect the arc AEB at E.

For, draw the radii AD and BD. Then, in the two triangles ACD, CBD, we have

$$AC=CB, \qquad AD=BD$$

and CD common: hence, the two triangles have their corresponding angles equal (Bk I. Th. viii), and consequently, ACD is equal to BCD. But since ACD is equal to BCD, it follows that the arc AE, which measures the former, is equal to the arc BE. which measures the latter.

PROBLEM VI.

At a given point in a straight line to erect a perpendicular to the line.

Let A be the given point, and BC the given line.

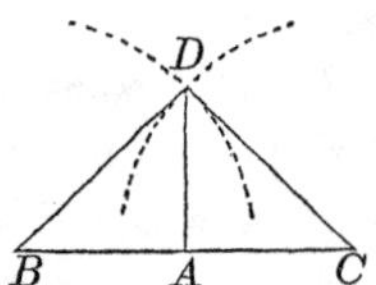

From A lay off any two distances, AB and AC, equal to each other. Then, from the points B and C, as centres, with a radius greater than AB, describe two arcs intersecting each other at D: draw DA, and it will be the perpendicular required.

For, draw the equal radii BD, DC. Then, the two triangles, BDA, and CDA, will have

$$AB=AC \qquad BD=DC$$

and AD common: hence, the angle DAB is equal to the angle DAC (Bk. I. Th. viii), and consequently, DA is perpendicular to BC.

SECOND METHOD.

When the point A is near the extremity of the line.

Assume any centre, as P, out of the given line. Then with P as a centre, and radius from P to A, describe the circumference of a circle. Through C, where the circumference cuts BA, draw CPD. Then, through D, where CP produced meets the circumference, draw DA: then will DA be perpendicular to BA, since CAD is an angle in a semicircle (Bk. II. Th. x).

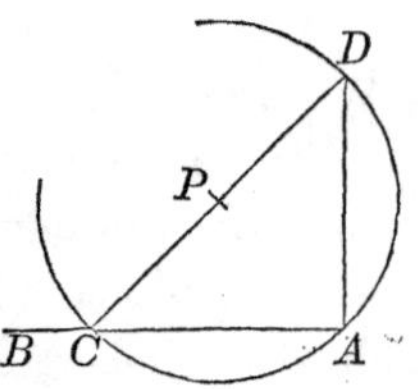

PROBLEM VII.

From a given point without a straight line to let fall a perpendicular on the line.

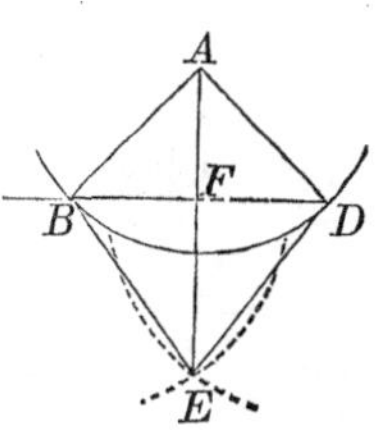

Let A be the given point, and BD the given line.

From the point A as a centre, with a radius greater than the shortest distance to BD, describe an arc cutting BD in the points B and D. Then, with B and D as centres, and the same radius, describe two arcs intersecting each other at E. Draw AFE, and it will be the perpendicular required.

For, draw the equal radii AB, AD, BE and DE. Then, the two triangles EAB and EAD will have the sides of the one equal to the sides of the other, each to each; hence, their corresponding angles will be equal (Bk. I. Th. viii), viz. the angle BAE to the angle DAE. Hence, the two triangles BAF and DAF will have two sides and the included angle of the one, equal to two sides and the included angle of the other, and therefore, the angle AFB will be equal to the angle AFD (Bk. I. Th. iv): hence, AFE will be perpendicular to BD.

SECOND METHOD.

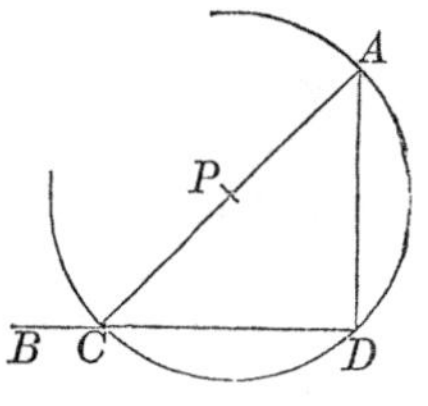

When the given point A is nearly opposite the extremity of the line.

Draw AC, to any point C of the line BD. Bisect AC at P. Then, with P as a centre and PC as a radius, describe the semicircle CDA; draw AD, and it will be perpendicular to CD, since CDA is an angle in a semicircle (Bk. II. Th. x).

PROBLEM VIII.

At a given point in a given line, to make an angle equal to a given angle.

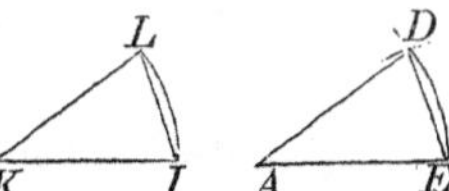

Let A be the given point, AE the given line, and IKL the given angle.

From the vertex K, as a centre, with any radius, describe the arc IL, terminating in the two sides of the angle: and draw the chord IL.

From the point A, as a centre, with a distance AE, equal to KI, describe the arc DE; then with E, as a centre, and a radius equal to the chord IL, describe an arc cutting DE at D; draw AD, and the angle EAD will be equal to the angle K.

For, draw the chord DE. Then the two triangles IKL and EAD, having the three sides of the one equal to the three sides of the other, each to each, the angle EAD will be equal to the angle K (Bk. I. Th. viii).

PROBLEM IX.

Through a given point to draw a line that shall be parallel to a given line.

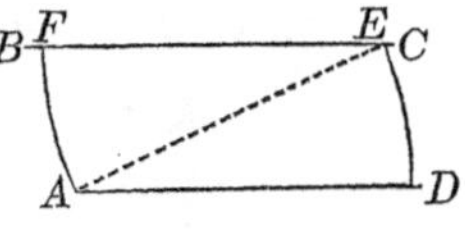

Let A be the given point and BC the given line.

With A as a centre, and any radius greater than the shortest distance from A to BC, describe the indefinite arc DE. From the point E, as a centre, with the same radius, describe the arc AF: then, make ED equal to AF and draw AD, and it will be the required parallel.

For, since the arcs AF and ED are equal, the angles EAD and AEF, which they measure, are qual: hence, the line AD is parallel to BC (Bk I. Th. xiii).

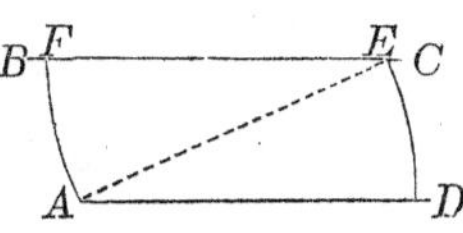

PROBLEM X.

Two angles of a triangle being given or known, to find the third.

Draw the indefinite line DEF.

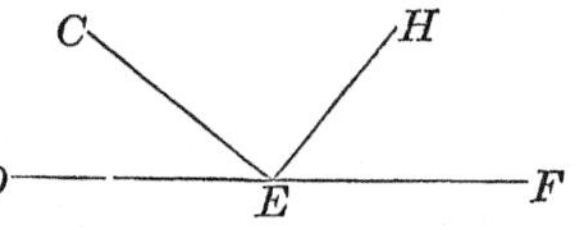

At any point, as E, make the angle DEC equal to one of the given angles, and then CEH equal to a second, by Prob. VIII; then will the angle HEF be equal to the third angle of the triangle.

For, the sum of the three angles of a triangle is equal to two right angles (Bk. I. Th. xvii); and the sum of the three angles on the same side of the line DE is equal to two right angles (Bk. I. Th. ii. Cor. 2); hence, if DEC and CEH are equal to two of the angles, the angle HEF will be equal to the remaining angle of the triangle.

PROBLEM XI.

Three sides of a triangle being given, to describe the triangle.

Let A, B, and C, be the given ides.

Draw DE, and make it equal to the side A. From the point D, as a centre, with a radius equal to the second side B, describe an arc:

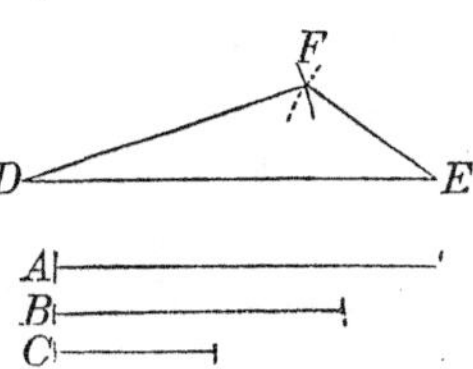

from E as a centre, with the third side C, describe another arc intersecting the former in F: draw DF and FE: then will DEF be the required triangle.

For, the three sides are respectively equal to the three lines A, B, and C.

PROBLEM XII.

The adjacent sides of a parallelogram, with the angle which they contain, being given, to describe the parallelogram.

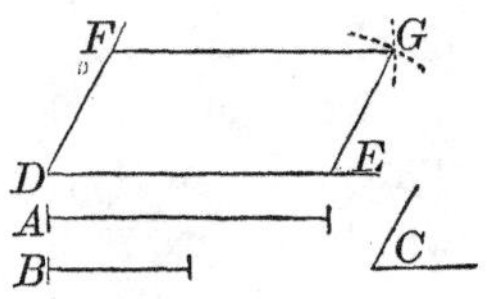

Let A and B be the given sides and C the given angle.

Draw the line DE and make it equal to A. At the point D make the angle EDF equal to the angle C. Make the side DF equal to B. Then describe two arcs, one from F, as a centre, with a radius FG equal to DE, the other from E, as a centre, with a radius EG equal to DF. Through the point G, the point of intersection, draw the lines EG and FG, and $DEGF$ will be the required parallelogram.

For, in the quadrilateral $DFGE$, the opposite sides DE and FG are each equal to A: the opposite sides DF and EG are each equal to B, and the angle EDF is equal to C. But, since the opposite sides are equal, they are also parallel (Bk. I. Th. xxiv), and therefore the figure is parallelogram.

PROBLEM XIII.

To describe a square on a given line.

Let AB be the given line.

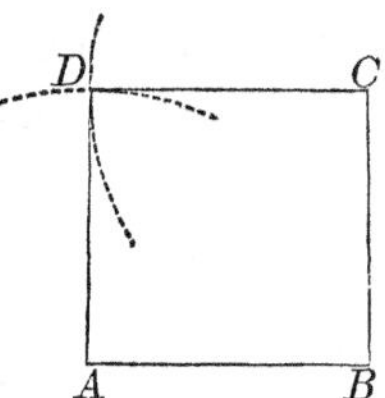

At the point B draw BC perpendicular to AB, by Problem VI, and then make it equal to AB.

Then, with A as a centre, and radius equal to AB, describe an arc; and with C as a centre, and the same radius AB, describe another arc; and through D, their point of intersection, draw AD and CD: then will $ABCD$ be the required square.

For, since the opposite sides are equal, the figure will be a parallelogram (Bk. I. Th. xxiv): and since one of the angles is a right angle, the others will also be right angles (Bk. I. Th. xxiii. Cor. 1); and since the sides are all equal, the figure will be a square.

PROBLEM XIV.

To construct a rhombus, having given the length of one of the equal sides, and one of the angles.

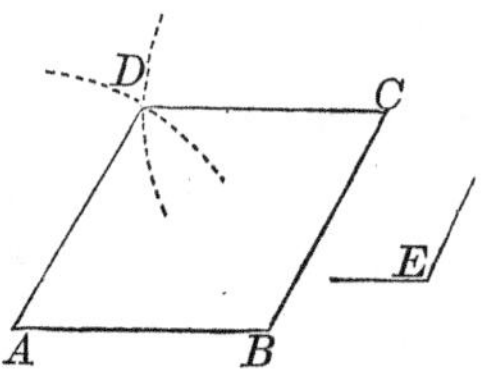

Let AB be equal to the given side, and E the given angle.

At B lay off an angle, ABC, equal to E, by Prob. VIII. and make BC equal to AB. Then, with A and C as centres, and a radius equal to AB, describe two arcs. Through D, their point of intersection, draw the lines AD, CD: then will $ABCD$ be the required rhombus.

For, since the opposite sides are equal, they will be parallel (Bk. I. Th. xxiv). But they are each equal to AB, and the angle B is equal to the angle E: hence, $ABCD$ is the required rhombus.

PROBLEM XV.

To find the centre of a circle.

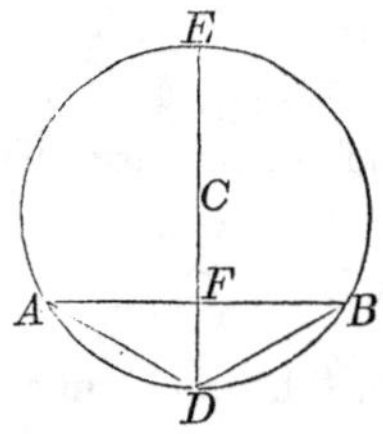

Draw any chord, as AB, and bisect it by Problem IV. Then, through F, the middle point, draw DCE, perpendicular to AB, by Problem VI. Then DCE will be a diameter of the circle (Bk. II. Th. ii. Cor.). Then bisect DE at C, and C will be the centre of the circle.

PROBLEM XVI.

To describe the circumference of a circle through three given points.

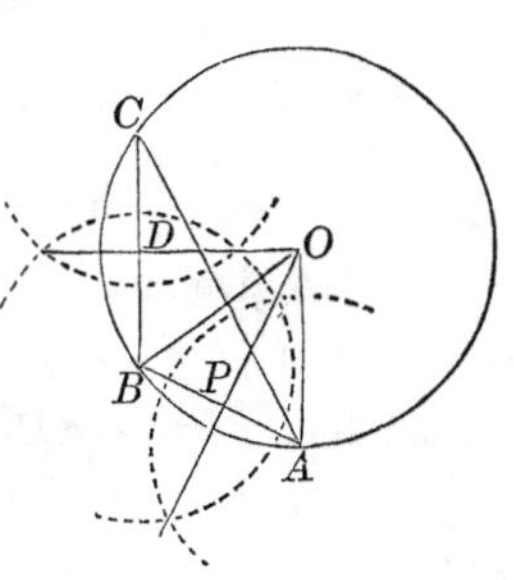

Let A, B, C, be the given points. Join these points by the straight lines AC, AB, BC.

Then, bisect any two of these straight lines, as AB, BC, by the perpendiculars OD, OP (Prob. iv); and the point O, where these perpendiculars intersect each other, will be the centre of the circle.

Then with O as a centre, and a radius equal to OA, describe the circumference of a circle, and it will pass through the points A, B, and C.

For, the two right angled triangles OAP and OBP have the side AP equal to the side BP, OP common, and the included angles OPA and OPB equal, being right angles; hence, the side OB is equal to OA (Bk. I. Th. iv).

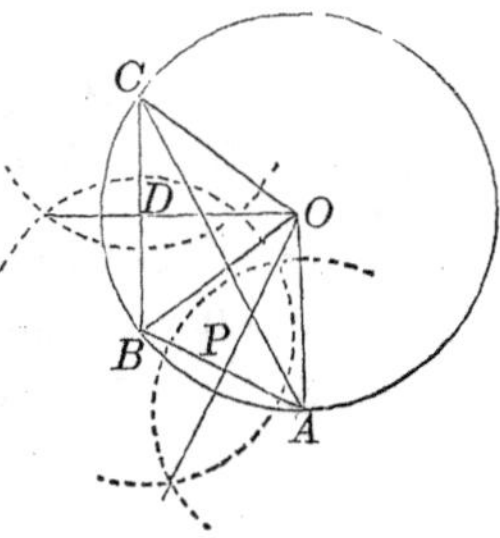

In like manner it may be shown, that OC is equal to OB. Hence, a circumference described with the radius OA, will pass through the points B and C.

Sch. This problem enables us to describe the circumference of a circle about a given triangle. For, we may consider the vertices of the three angles as the three points through which the circumference is to pass.

PROBLEM XVII.

Through a given point in the circumference of a circle, to draw a tangent line to the circle.

Let A be the given point.

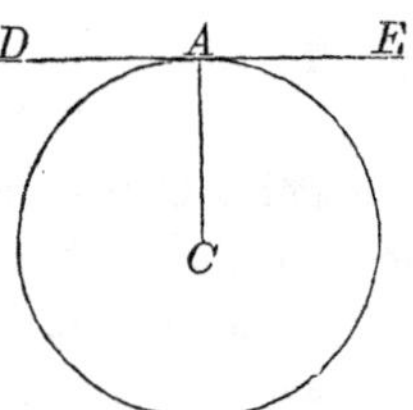

Through A, draw the radius AC to the centre, and then draw DAE perpendicular to AC, by Problem VI. Then will DAE be tangent to the circle at the point A (Bk. II. Th. v).

PROBLEM XVIII.

Through a given point without the circumference, to draw a tangent line to the circle.

Let C be the centre of the circle, and A the given point without the circle.

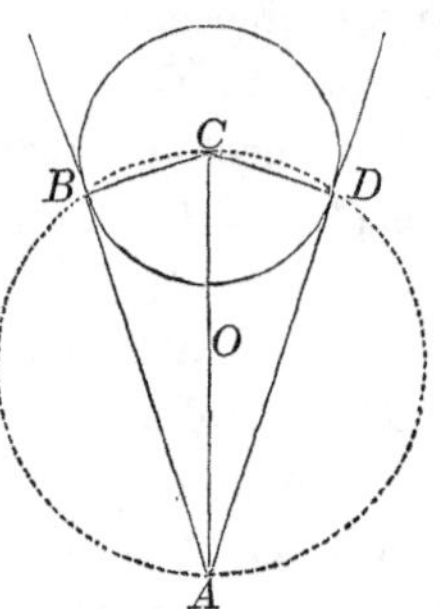

Join A and the centre C, and on AC, as a diameter, describe a circumference. Through the points B and D, where the two circumferences intersect each other, draw the lines AB and AD: these lines will be tangent to the circle whose centre is C.

For, since the angles ABC and ADC are each inscribed in a semicircle, they will be right angles (Bk. II. Th. x). Again, since the lines AB, AD, are each perpendicular to a radius at its extremity, they will be tangent to the circle (Bk. II. Th. v).

PROBLEM XIX.

To inscribe a circle in a given triangle.

Let ABC be the given triangle.

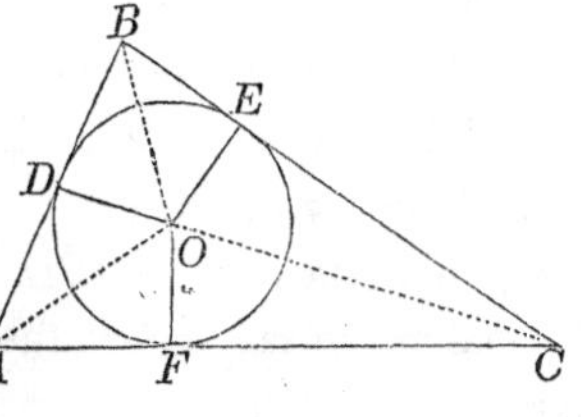

Bisect the angles A and B by the lines AO and BO, meeting at the point O. From O, let fall the perpendiculars OD, OE, OF, on the three sides of the triangle—these perpendiculars will be equal to each other.

For, in the two right angled triangles DAO and FAO, we have the right angle D equal the right angle F, the angle FAO equal to DAO, and consequently, the third angles AOD and AOF are equal (Bk. I. Th. xvii. Cor 1). But the two triangles have a common side AO, hence, they are equal (Bk. I. Th. v), and consequently, OD is equal to OF.

In a similar manner, it may be proved that OE and OD are equal: hence, the three perpendiculars, OD, OF, and OE, are all equal.

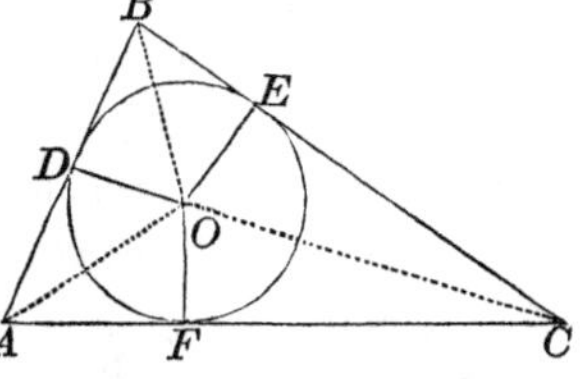

Now, if with O as a centre, and OF as a radius, we describe the circumference of a circle, it will pass through the points D and E. and since the sides of the triangle are perpendicular to the radii OF, OD, OE, they will be tangent to the circumference (Bk. II. Th. v). Hence, the circle will be inscribed in the triangle.

PROBLEM XX.

To inscribe an equilateral triangle in a circle.

Through the centre C draw any diameter, as ACB. From B as a centre, with a radius equal to BC, describe the arc DCE. Then, draw AD, AE, and DE, and DAE will be the required triangle.

For, since the chords BD, BE, are each equal to the radius CB, the arcs BD, BE, are each equal to sixty degrees (Bk. II. Th. XIX), and the arc DBE to one hundred and twenty degrees; hence, the angle DAE is equal to sixty degrees (Bk. II. Th. viii).

Again, since the arc BD is equal to sixty degrees, and the arc BDA equal to one hundred and eighty degrees, it follows that DA will be equal to one hundred and twenty degrees: hence, the angle DEA is equal to sixty degrees, and consequently, the third angle ADE, is equal to sixty degrees.

Therefore, the triangle *ADE* is equilateral (Bk. I. Th. vi. Cor. 2).

PROBLEM XXI.

To inscribe a regular hexagon in a circle.

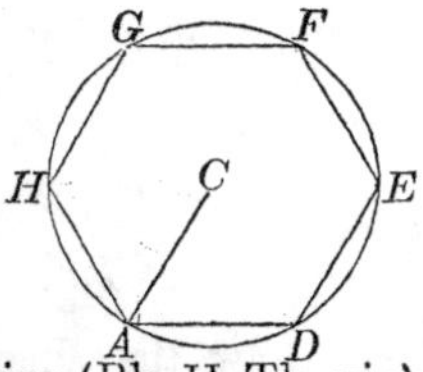

Draw any radius, as *AC*. Then apply the radius *AC* around the circumference, and it will give the chords *AD*, *DE*, *EF*, *FG*, *GH*, and *HA*, which will be the sides of the regular hexagon. For, the side of a hexagon is equal to the radius (Bk. II. Th. xix)

PROBLEM XXII.

To inscribe a square in a given circle.

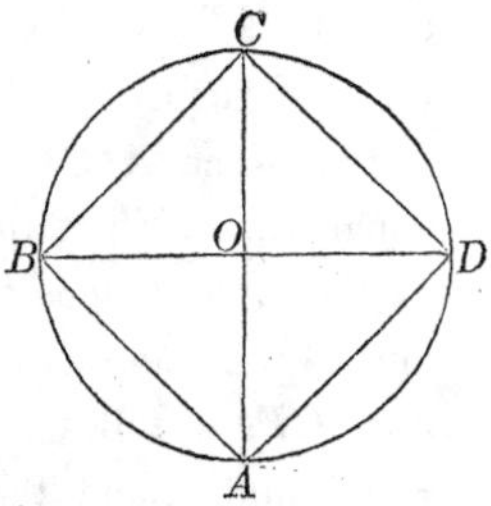

Let *ABCD* be the given circle. Draw the two diameters *AC*, *BD*, at right angles to each other, and through the points *A*, *B*, *C* and *D* draw the lines *AB*, *BC*, *CD*, and *DA*: then will *ABCD* be the required square.

For, the four right angled triangles, *AOB*, *BOC*, *COD*, and *DOA* are equal, since the sides *AO*, *OB*, *OC*, and *OD* are equal, being radii of the circle; and the angles at *O* are equal in each, being right angles: hence, the sides *AB*, *BC*, *CD*, and *DA* are equal (Bk. I. Th. iv).

But each of the angles *ABC*, *BCD*, *CDA*, *DAB*, is a right angle, being an angle in a semicircle (Bk. II. Th. x): hence, the figure *ABCD* is a square (Bk. I. Def. 48)

Sch. If we bisect the arcs AB, BC, CD, DA, and join the points, we shall have a regular octagon inscribed in the circle. If we again bisect the arcs, and join the points of bisection, we shall have a regular polygon of sixteen sides.

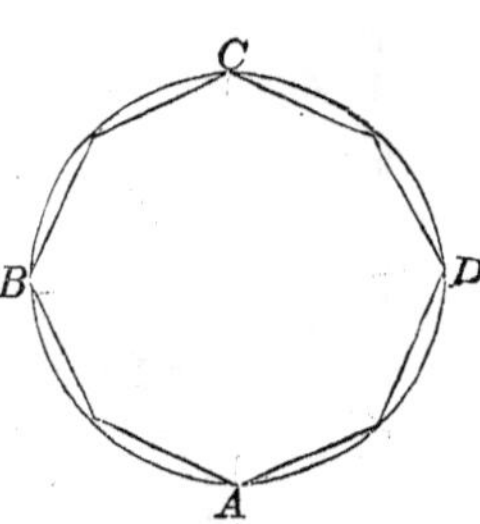

PROBLEM XXIII.

To describe a square about a given circle.

Draw the diameters AB, DE, at right angles to each other. Through the extremities A and B draw FAG and HBI parallel to DE, and through E and D, draw FEH and GDI parallel to AB: then will $FGIH$ be the required square.

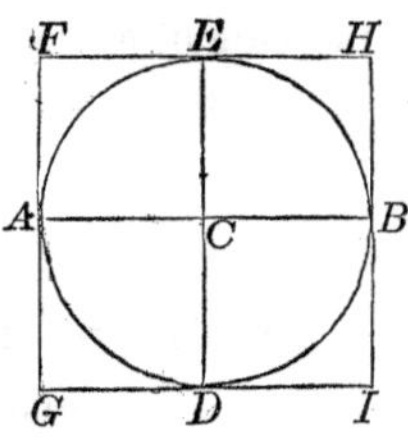

For, since $ACDG$ is a parallelogram, the opposite sides are equal (Bk. I. Th. xxiii): and since the angle at C is a right angle, all the other angles are right angles (Bk. I. Th. xxiii. Cor. 1): and as the same may be proved of each of the figures CI, CH and CF, it follows that all the angles, F, G, I, and H, are right angles, and that the sides GI, IH, HF, and FG, are equal, each being equal to the diameter of the circle. Hence, he figure $GIHF$ is a square (Bk. I. Def. 48).

GEOMETRY.

BOOK III.

OF RATIOS AND PROPORTIONS.

DEFINITIONS.

1. *Ratio* is the quotient arising from dividing one quantity by another quantity of the same kind. Thus, if the numbers 3 and 6 have the same unit, the ratio of 3 to 6 will be expressed by

$$\frac{6}{3}=2.$$

And in general, if A and B represent quantities of the same kind, the ratio of A to B will be expressed by

$$\frac{A}{B}.$$

2. If there be four numbers, 2, 4, 8, 16, having such values that the second divided by the first is equal to the fourth divided by the third, the numbers are said to be in proportion. And in general, if there be four quantities, A, B, C, and D, having such values that

$$\frac{B}{A}=\frac{D}{C},$$

then, A is said to have the same *ratio* to B, that C has to D; or, the ratio of A to B is equal to the ratio of C to D When four quantities have this relation to each other, they are said to be in proportion. Hence, *proportion is an equality of ratios.*

To express that the ratio of A to B is equal to the ratio of C to D, we write the quantities thus:

$$A : B :: C : D;$$

and read, A is to B, as C to D.

The quantities which are compared together are called the *terms* of the proportion. The first and last terms are called the *two extremes*, and the second and third terms, the *two means*. Thus, A and D are the extremes, and B and C the means.

3. Of four proportional quantities, the first and third are called the *antecedents*, and the second and fourth the *consequents;* and the last is said to be a fourth proportional to the other three taken in order. Thus, in the last proportion, A and C are the antecedents, and B and D the consequents.

4. Three quantities are in proportion when the first has the same ratio to the second, that the second has to the third; and then the middle term is said to be a mean proportional between the other two. For example,

$$3 : 6 :: 6 : 12;$$

and 6 is a mean proportional between 3 and 12.

5. Quantities are said to be in proportion by *inversion*, or *inversely*, when the consequents are made the antecedents and the antecedents the consequents.

Thus, if we have the proportion

$$3 : 6 :: 8 : 16.$$

the inverse proportion would be

$$6 : 3 :: 16 : 8.$$

6. Quantities are said to be in proportion by *alternation*, or *alternately*, when antecedent is compared with antecedent and consequent with consequent.

Thus, if we have the proportion

$$3 : 6 :: 8 : 16,$$

the alternate proportion would be

$$3 : 8 :: 6 : 16.$$

7. Quantities are said to be in proportion by *composition*, when the sum of the antecedent and consequent is compared either with antecedent or consequent.

Thus, if we have the proportion

$$2 : 4 :: 8 : 16,$$

the proportion by composition would be

$$2+4 : 4 :: 8+16 : 16;$$

that is, $6 : 4 :: 24 : 16.$

8. Quantities are said to be in proportion by *division*, when the difference of the antecedent and consequent is compared either with the antecedent or consequent.

Thus, if we have the proportion

$$3 : 9 :: 12 : 36,$$

the proportion by division will be

$$9-3 : 9 :: 36-12 : 36;$$

that is, $6 : 9 :: 24 : 36.$

9. Equimultiples of two or more quantities are the products which arise from multiplying the quantities by the same number.

Thus, if we have any two numbers, as 6 and 5, and multiply

them both by any number, as 9, the equimultiples will be 54 and 45 ; for

$$6 \times 9 = 54 \qquad \text{and} \qquad 5 \times 9 = 45.$$

Also, $m \times A$ and $m \times B$ are equimultiples of A and B, the common multiplier being m.

10. Two quantities, A and B, are said to be *reciprocally proportional*, or *inversely proportional*, when one increases in the same ratio as the other diminishes. When this relation exists, either of them is equal to a constant quantity divided by the other.

Thus, if we had any two numbers, as 2 and 4, so related to each other that if we divided one by any number we must multiply the other by the same number, one would increase just as fast as the other would diminish, and their product would not be changed.

THEOREM I.

If four quantities are in proportion, the product of the two extremes will be equal to the product of the two means.

If we have the proportion

$$A : B :: C : D$$

we have, by Def. 2,

$$\frac{B}{A} = \frac{D}{C}$$

and by clearing the equation of fractions, we have

$$BC = AD$$

Sch. The general principle is verified in the proportion between the numbers

$$2 : 10 :: 12 : 60$$

which gives

$$2 \times 60 = 10 \times 12 = 120$$

THEOREM II.

If four quantities are so related to each other, that the product of two of them is equal to the product of the other two; then, two of them may be made the means, and the other two the extremes of a proportion.

Let A, B, C, and D, have such values that

$$B \times C = A \times D$$

Divide both sides of the equation by A, and we have

$$\frac{B}{A} \times C = D$$

Then divide both sides of the last equation by C, and we have

$$\frac{B}{A} = \frac{D}{C}$$

hence, by Def. 2, we have

$$A \ : \ B \ :: \ C \ : \ D.$$

Sch. The general truth may be verified by the numbers

$$2 \times 18 = 9 \times 4$$

which give

$$2 \ : \ 4 \ :: \ 9 \ : \ 18$$

THEOREM III.

If three quantities are in proportion, the product of the two extremes will be equal to the square of the middle term.

Let us suppose that we have

$$A \ : \ B \ :: \ B \ : \ C$$

Then, by Def. 2, we have

$$\frac{B}{A} = \frac{C}{B}$$

and by clearing the equation of its fractions, we have

$$B^2 = C \times A$$

Sch. The proposition may be verified by the numbers

$$3 : 6 :: 6 : 12$$

which give

$$3 \times 12 = 6 \times 6 = 36$$

THEOREM IV.

If four quantities are in proportion, they will be in proportion when taken alternately.

Let $A : B :: C : D$

Then, by Def. 2, we have

$$\frac{B}{A} = \frac{D}{C}$$

Multiplying both members of this equation by $\frac{C}{B}$, we have

$$\frac{C}{A} = \frac{D}{B}$$

and consequently,

$$A : C :: B : D.$$

Sch. The theorem may be verified by the proportion

$$10 : 15 :: 20 : 30$$

for, we have, by alternation,

$$10 : 20 :: 15 : 30.$$

THEOREM V.

If there be two sets of proportions, having an antecedent and a consequent in the one, equal to an antecedent and a consequent in the other; then, the remaining terms will be proportional.

If we have

$A . B :: C : D$, and $A : B :: E : F$;

then we shall have

$$\frac{B}{A}=\frac{D}{C} \quad \text{and} \quad \frac{B}{A}=\frac{F}{E}$$

Hence, by Ax. 1, we have

$$\frac{D}{C}=\frac{F}{E}$$

and consequently,

$$C : D :: E : F.$$

Sch The proposition may be verified by the following proportions,

2 : 6 :: 8 : 24 and 2 : 6 :: 10 : 30

which give

8 : 24 :: 10 : 30.

THEOREM VI.

If four quantities are in proportion, they will be in proportion when taken inversely.

If we have the proportion

$$A : B :: C : D$$

we have, by Th. I,

$$A \times D = B \times C,$$

or

$$B \times C = A \times D.$$

Hence, we have, by Th. II,

$$B : A :: D : C.$$

Sch. The proposition may be verified by the proportion

7 : 14 :: 8 : 16;

which, when taken inversely, gives

14 : 7 :: 16 : 8.

THEOREM VII.

If four quantities are in proportion, they will be in proportion by composition

Let us suppose that we have

$$A : B :: C : D$$

we shall then have

$$A \times D = B \times C.$$

To each of these equals, add $B \times D$, and we have

$$(A+B) \times D = (C+D) \times B;$$

and by separating the factors by Th. II, we have

$$A+B : B :: C+D : D.$$

Sch. The proposition may be verified by the following proportion,

$$9 : 27 :: 16 : 48.$$

We shall have, by composition,

$$9+27 : 27 :: 16+48 : 48,$$

that is, $36 : 27 :: 64 : 48,$

in which the ratio is three fourths.

THEOREM VIII.

If four quantities are in proportion, they will be in proportion by division.

Let us suppose that we have

$$A : B :: C : D;$$

we shall then have

$$A \times D = B \times C.$$

From each of these equals let us subtract $B \times D$, and we have

$$(A-B) \times D = (C-D) \times B;$$

and by separating the factors by Th. II, we have,

$$A-B : B :: C-D : D.$$

Sch. The proposition may be verified by the proportion,

$$24 : 8 :: 48 : 16.$$

We have, by division,

$$24-8 \ : \ 8 \ :: \ 48-16 \ : \ 16;$$

that is, $16 \ : \ 8 \ :: \ 32 \ : \ 16;$

in which the ratio is one-half.

THEOREM IX.

Equal multiples of two quantities have the same ratio as the quantities themselves.

If we have the proportion

$$A \ : \ B \ :: \ C \ : \ D$$

we shall have

$$\frac{B}{A}=\frac{D}{C}$$

Now, let M be any number, and by it multiply the numerator and denominator of the first member of the equation which will not change its value: we. shall then have

$$\frac{M\times B}{M\times A}=\frac{D}{C}$$

and hence we have

$$M\times A \ : \ M\times B \ :: \ C \ : \ D,$$

that is, the equal multipliers $M\times A$ and $M\times B$, have the same ratio as A to B.

Sch. The proposition may be verified by the proportion,

$$5 \ : \ 10 \ :: \ 12 \ : \ 24;$$

for, by multiplying the first antecedent and consequent by any number, as 6, we have

$$30 \ : \ 60 \ :: \ 12 \ : \ 24,$$

in which the ratio is still 2.

THEOREM X.

If four quantities are proportional, and one antecedent and its consequent be augmented by quantities which have the same ratio as the antecedent and consequent, the four quantities will still be in proportion.

Let us take the proportions

$A : B :: C : D$, and $A : B :: E : F$,
which give

$$A \times D = B \times C \qquad \text{and} \qquad A \times F = B \times E;$$

adding these equals we have

$$A \times (D+F) = B \times (C+E);$$

and by Th. II, we have

$$A : B :: C+E : D+F$$

in which the antecedent C and its consequent D, are augmented by the quantities E and F, which have the same ratio.

Sch. The proposition may be verified by the proportion,

$$9 : 18 :: 20 : 40,$$

in which the ratio is 2.

If we augment the antecedent and its consequent by 15 and 30, which have the same ratio, we have

$$9 : 18 :: 20+15 : 40+30$$

that is, $9 : 18 :: 35 : 70$,
in which the ratio is still 2.

THEOREM XI.

If four quantities are proportional, and one antecedent and its consequent be diminished by quantities which have the same ratio as the antecedent and consequent, the four quantities will still be in proportion.

Let us take the proportions

$A : B :: C : D$, and $A : B :: E : F$,

which give

$$A \times D = B \times C \quad \text{and} \quad A \times F = B \times E.$$

By subtracting these equalities, we have

$$A \times (D - F) = B \times (C - E);$$

and by Th. II, we obtain

$$A : B :: C - E : D - F,$$

in which the antecedent and consequent, C and D, are diminished by E and F, which have the same ratio.

Sch. The proposition may be verified by the proportion,

$$9 : 18 :: 20 : 40,$$

for, by diminishing the antecedent and consequent by 15 and 30, we have

$$9 : 18 :: 20 - 15 : 40 - 30;$$

that is $\quad 9 : 18 :: 5 : 10$

in which the ratio is still 2.

THEOREM XII.

If we have several sets of proportions, having the same ratio, any antecedent will be to its consequent, as the sum of the antecedents to the sum of the consequents.

If we have the several proportions,

$A : B :: C : D$ which gives $A \times D = B \times C$

$A : B :: E : F$ which gives $A \times F = B \times E$

$A : B :: G : H$ which gives $A \times H = B \times G$

We shall then have, by addition,

$$A \times (D + F + H) = B \times (C + E + G);$$

and consequently, by Th. II.

$$A : B :: C + E + G : D + F + H.$$

Sch. The proposition may be verified by the following proportions: viz.

2 : 4 :: 6 : 12 and 1 : 2 :: 3 : 6

Then, 2 : 4 :: 6+3 : 12+6;

hat is, 2 : 4 :: 9 : 18,

in which the ratio is still 2.

THEOREM XIII.

If four quantities are in proportion, their squares or cubes will also be proportional.

If we have the proportion

$$A : B :: C : D,$$

it gives

$$\frac{B}{A}=\frac{D}{C}$$

Then, if we square both members, we have

$$\frac{B^2}{A^2}=\frac{D^2}{C^2}$$

and if we cube both members, we have

$$\frac{B^3}{A^3}=\frac{D^3}{C^3}$$

and then, changing these equalities into a proportion, we have for the first,

$$A^2 : B^2 :: C^2 : D^2;$$

and for the second

$$A^3 : B^3 :: C^3 : D^3.$$

Sch. We may verify the proposition by the proportion,

2 : 4 :: 6 : 12,

and by squaring each term we have,

4 : 16 :: 36 : 144

numbers which are still proportional, and in which the ratio is 4.

If we cube the numbers we have,

$$2^3 : 4^3 :: 6^3 : 12^3$$

that is, $8 : 64 :: 216 : 1728$,

in which the ratio is 8.

THEOREM XIV.

If we have two sets of proportional quantities, the products of the corresponding terms will be proportional.

Let us take the proportions,

$$A : B :: C : D \quad \text{which gives} \quad \frac{B}{A}=\frac{D}{C}$$

$$E : F :: G : H \quad \text{which gives} \quad \frac{F}{E}=\frac{H}{G}$$

Multiplying the equalities together, we have

$$\frac{B \times F}{A \times E}=\frac{D \times H}{C \times G}$$

and this by Th. II, gives

$$A \times E : B \times F :: C \times G : D \times H.$$

Sch. The proposition may be verified by the following proportions:

$$8 : 12 :: 10 : 15,$$

and $3 : 4 :: 6 : 8$;

we shall then have

$$24 : 48 :: 60 : 120$$

which are proportional, the ratio being 2.

GEOMETRY.

BOOK IV

OF THE MEASUREMENT OF AREAS, AND THE PROPORTIONS OF FIGURES.

DEFINITIONS.

1. Similar figures, are those which have the angles of the one equal to the angles of the other, each to each, and the sides about the equal angles proportional.

2. Any two sides, or any two angles, which are like placed in the two similar figures, are called *homologous* sides or angles.

3. A polygon which has all its angles equal, each to each, and all its sides equal, each to each, is called a *regular polygon* A regular polygon is both equiangular and equilateral.

4. If the length of a line be computed in feet, one foot is the unit of the line, and is called the *linear unit.* If the length of a line be computed in yards, one yard is the linear unit.

5. If we describe a square on the unit of length, such square is called the unit of surface. Thus, if the linear unit is one foot, one square foot will be the unit of surface.

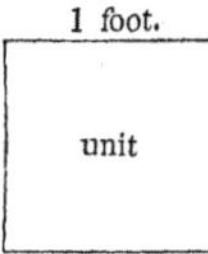

6. If the linear unit is one yard, one square yard will be the unit of surface; and this square yard contains nine square feet.

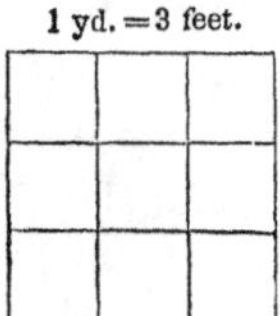

7. The *area* of a figure is the measure of its surface. The unit of the number which expresses the area, is a square, the side of which is the unit of length.

8. Figures have equal areas, when they contain the same measuring unit an equal number of times.

9. Figures which have equal areas are called *equivalent.* The term *equal,* when applied to figures, implies an equality in all respects. Such figures being applied to each other, will coincide in all their parts. The term *equivalent,* implies an equality in one respect only: viz. an equality in their areas

THEOREM I.

Parallelograms which have equal bases and equal altitudes, are equivalent.

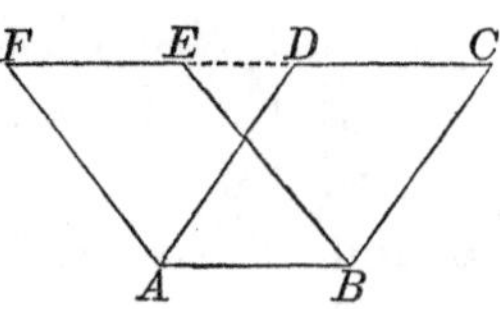

Place the base of one parallelogram on that of the other, so that AB shall be the common base of the two parallelograms $ABCD$ and $ABEF$. Now, since the parallelograms have the same altitude, their upper bases, DC and FE, will fall on the same line $FEDC$, parallel to AB. Since the opposite sides of a parallelogram are equal to each other (Bk. I. Th. xxiii), AD is equal to BC. Also, DC and FE are each equal to AB: and consequently, they are equal to each other (Ax. 1). To each, add ED: then will CE be equal to DF.

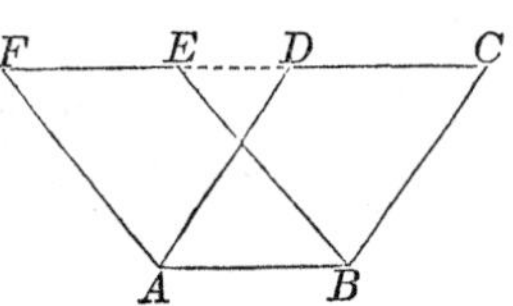

But since the line FC cuts the wo parallels CB and DA, the angle BCE will be equal to the angle ADF (Bk. I. Th. xiv): hence, the two triangles ADF and BCE have two sides and the included angle of the one equal to two sides and the included angle of the other, each to each; consequently, they are equal (Bk. I. Th. iv).

If then, from the whole space $ABCF$ we take away the triangle ADF, there will remain the parallellogram $ABCD$; but if we take away the equal triangle BEC, there will remain the parallelogram $ABEF$: hence, the parallelogram $ABEF$ is equivalent to the parallelogram $ABCD$ (Ax. 3).

Cor. A parallelogram and a rectangle, having equal bases and equal altitudes, are equivalent.

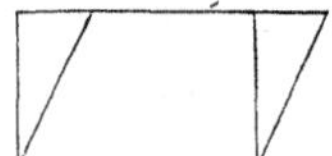

THEOREM II.

Triangles which have equal bases and equal altitudes, are equivalent.

Place the base of one triangle on that of the other, so that ABC and ABD shall be the two triangles, with the common base AB, and for their altitude the distance between the two parallels AB, FC: then will the triangle ABC be equivalent to the triangle ADB.

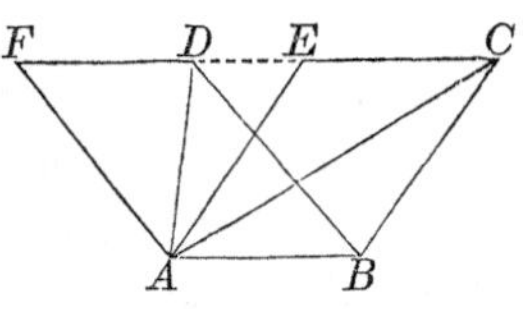

For, through A draw AE parallel to BC, and AF parallel to BD, forming the two parallelograms BE and BF. Then, since these parallelograms have a common base and equal altitudes, they will be equivalent (Th. i).

But the triangle ABC is half the parallelogram BE (Bk. I. Th. xxiii); and ABD is half the equal parallelogram BF: hence, the triangle ABC is equivalent to the triangle ABD.

THEOREM III.

If a triangle and a parallelogram have equal bases and equal altitudes, the triangle will be half the parallelogram.

Place the base of the triangle on the base of the parallelogram, so that AB shall be the common base of the triangle and parallelogram: then will the triangle ABE be half the parallelogram BD.

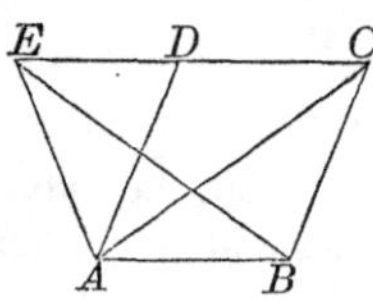

For, draw the diagonal AC. Then, since the altitude of the triangle AEB is equal to that of the parallelogram, the vertex will be found some where in CD, or in CD produced. Now the two triangles ABC and ABE, having the same base AB, and equal altitudes, are equivalent (Th. ii). But the triangle ABC is half the parallelogram BD (Bk. I. Th. xxiii): hence, the triangle ABE is half the parallelogram BD (Ax. 1).

Cor. Hence, if a triangle and a rectangle have equal bases and equal altitudes, the triangle will be half the rectangle.

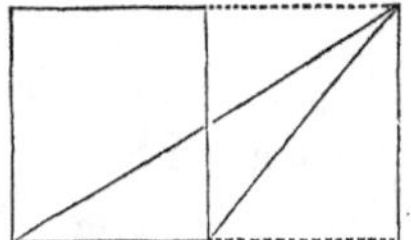

For, the rectangle would be equivalent to a parallelogram of the same base and altitude (Th. ix. Cor.), and since the triangle is half the parallelogram, it is also equivalent to half the rectangle.

THEOREM IV.

Rectangles which are described on equal lines are equivalent.

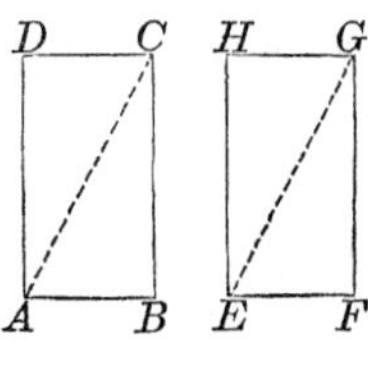

Let BD and FH be two rectangles, having the sides AB, BC, equal to the two sides EF, FG, each to each: then will the rectangle $ABCD$, described on the lines AB, BC, be equivalent to the rectangle $EFGH$, described on the lines EF, FG.

For, draw the diagonals AC, EG, dividing each parallelogram into two equal parts.

Then the two triangles, ABC, EFG, having two sides and the included angle of the one equal to two sides and the included angle of the other, each to each, are equal (Bk. I. Th. iv). But these equal triangles are halves of the respective rectangles (Th. iii. Cor.): hence, the rectangles are equal (Ax. 7); and consequently equivalent.

Cor. The squares on equal lines are equal. For a square is but a rectangle having its sides equal.

THEOREM V.

Two rectangles having equal altitudes are to each other as their bases.

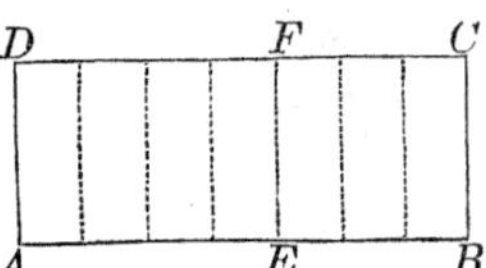

Let $AEFD$ and $EBCF$ be two rectangles having the common altitude AD; then will they be to each other as the bases AE and EB.

For, suppose the base AE to be to the base EB, as any two numbers, say the numbers 4 and 3. Let AE be then divided into four equal parts, and EB into three equal parts, and through the points of division draw parallels to AD. We shall thus form seven rectangles, all equivalent to each other since they have equal bases and equal altitudes (Th. iv).

But the rectangle $AEFD$ will contain four of these partial rectangles, while the rectangle $EBCF$ will contain three; hence, the rectangle $AEFD$ will be to the rectangle $EBCF$ as 4 to 3; that is, as the base AE to the base EB.

The same reasoning may be applied to any other rectangles whose bases are whole numbers: hence,

$$AEFD \; : \; EBCF \; :: \; AE \; : \; EB.$$

THEOREM VI.

Any two rectangles are to each other as the products of their bases and altitudes.

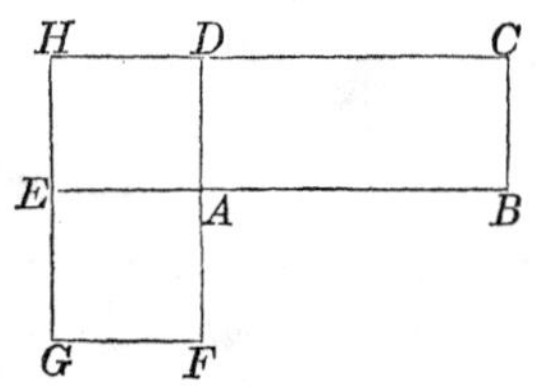

Let $ABCD$ and $AEGF$ be two rectangles: then will $ABCD : AEGF :: AB \times AD \cdot AF \times AE$

For, having placed the two rectangles so that BAE and DAF shall form straight lines, produce the sides CD and GE until they meet in H.

Then, the two rectangles $ABCD$, $AEHD$, having the common altitude AD, are to each other as their bases AB and AE (Th. v). In like manner, the two rectangles $AEHD$ $AEGF$, having the same altitude AE, are to each other as their bases AD and AF. Thus, we have the proportions

$$ABCD \; : \; AEHD \; :: \; AB \; : \; AE,$$
$$AEHD \; : \; AEGF \; :: \; AD \; : \; AF$$

If now, we multiply the corresponding terms together, the products will be proportional (Bk. III. Th. xiv); and the common multiplier $AEHD$ may be omitted (Bk. III. Th. ix): hence, we shall have

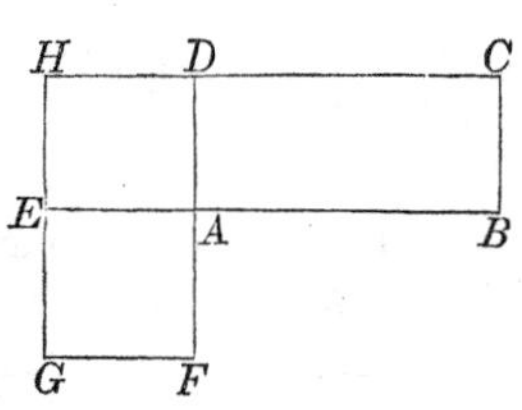

$$ABCD \ : \ AEGF \ :: \ AB \times AD \ : \ AE \times AF.$$

Sch. Hence, the product of the base by the altitude may be assumed as the measure of a rectangle. This product will give the number of superficial units in the surface: because, for one unit in height, there are as many superficial units as there are linear units in the base; for two units in height, twice as many; for three units in height three times as many, &c.

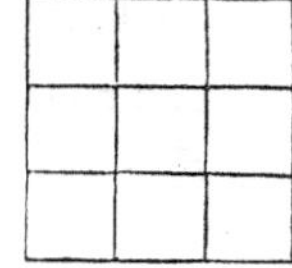

THEOREM VII.

The sum of the rectangles contained by one line, and the several parts of another line any way divided, is equivalent to the rectangle contained by the two whole lines.

Let AD be one line, and AB the other, divided into the parts AE, EF, FB: then will the rectangles contained by AD and AE, AD and EF, AD and FB, be equivalent to the rectangle AC which is contained by the lines AD and AB.

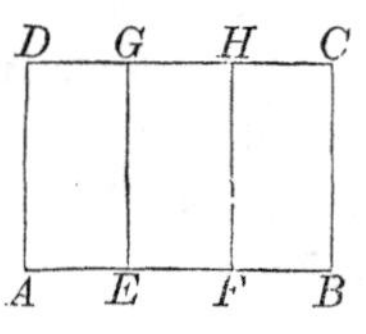

For, through E and F draw EG and FH parallel to AD, to which they will be equal (Bk. I. Th. xxiii). Then, AG will be equal to the rectangle of $AD \times AE$; EH will be equal to $EG \times EF$, or to $AD \times EF$; and FC will be equal to $FH \times FB$, or to $AD \times FB$.

But the rectangle AC is equal to the sum of the partial rectangles: hence,

$$AD \times AB = AD \times AE + AD \times EF + AD \times FB.$$

THEOREM VIII.

The area of any parallelogram is equal to the product of its base by its altitude.

Let $ABCD$ be any parallelogram, and BE its altitude: then will its area be equal to $AB \times BE$.

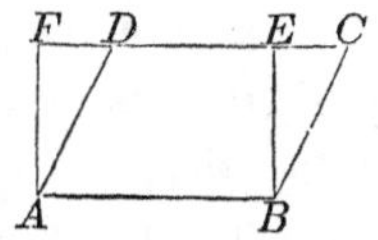

For, draw AF perpendicular to the base AB, and produce CD to F. Then, the parallelogram BD and the rectangle BF, having the same base and altitude are equivalent (Th. i. Cor.). But the area of the rectangle BF is equal to the product of its base AB by the altitude AF (Th. vi. Sch.): hence, the area of the parallelogram is equal to $AB \times BE$.

Cor. Parallelograms of equal bases are to each other as their altitudes; and if their altitudes are equal, they are to each other as their bases.

For, let B be the common base, and C and D the altitudes of two parallelograms. Then, by the theorem, their areas are to each other, as

$$B \times C \ : \ B \times D,$$

that is, (Bk. III. Th ix), as $\quad C : D.$

If A and B be their bases, and C their common altitude, then they will be to each other as

$A \times C \ : \ B \times C$: that is, as $\quad A : B.$

THEOREM IX.

The area of a triangle is equal to half the product of its base by its altitude.

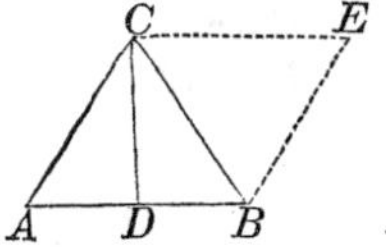

Let ABC be any triangle and CD its altitude: then will its area be equal to half the product of $AB \times CD$.

For, through B draw BE parallel to AC, and through C draw CE parallel to AB: we shall then form the parallelogram AE, having the same base and altitude as the triangle ABC.

But the area of the parallelogram is equal to the product of the base AB by its altitude DC; and since the parallelogram is double the triangle (Th. iii), it follows that the area of the triangle is equal to half this product: that is, to half the product of $AB \times CD$.

Cor. Two triangles of the same altitude are to each other as their bases; and two triangles of the same base are to each other as their altitudes. And generally, triangles are to each other as the products of their bases and altitudes.

THEOREM X.

The area of a trapezoid is equal to half the product of its altitude multiplied by the sum of its parallel sides.

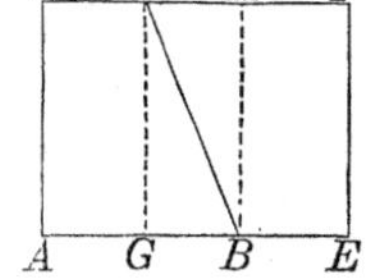

Let $ABCD$ be a trapezoid, CG its altitude, and AB, DC its parallel sides: then will its area be equal to half the product of

$$CG \times (AB + DC).$$

For, produce AB until BE is equal to DC, and complete the rectangle AF; also, draw BH perpendicular to AB.

Then, the rectangle AC will be equivalent to BF, since they have equal bases and equal altitudes (Th. iv). The diagonal BC will divide the rectangle GH into two equal triangles; and hence, the trapezoid $ABCD$ will be equivalent to the trapezoid $BEFC$; and consequently, the rectangle AF, is double the trapezoid $ABCD$.

But the rectangle AF is equivalent to the product of $AD \times AE$; that is, to $CG \times (AB+DC)$; and consequently, the trapezoid $ABCD$ is equal to half that product.

THEOREM XI.

If a line be divided into two parts, the square described on the whole line is equivalent to the sum of the squares described on the two parts, together with twice the rectangle contained by the parts.

Let the line AB be divided into two parts at the point E: then will the square described on AB be equivalent to the two squares described on AE and EB, together with twice the rectangle contained by AE and EB: that is

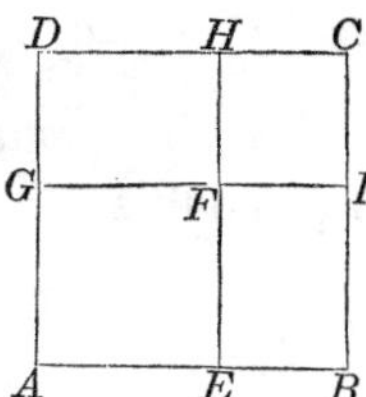

$$\overline{AB}^2 = \overline{AE}^2 + \overline{EB}^2 + 2AE \times EB.$$

For, let AC be a square on AB, and AF a square on AE, and produce the sides EF and GF to H and I.

Then, since EH is equal to AD, being the opposite side of a rectangle, it is also equal to AB; and GI is likewise equal to AB. If, therefore, from these equals we take away EF and

GF, there will remain FH equal to FI, and each will be equal to HC or IC; and since the angle at F is a right angle, it follows that FC is equal to a square described on EB. It also follows, that DF and FB are each equal to the rectangle of AE into EB.

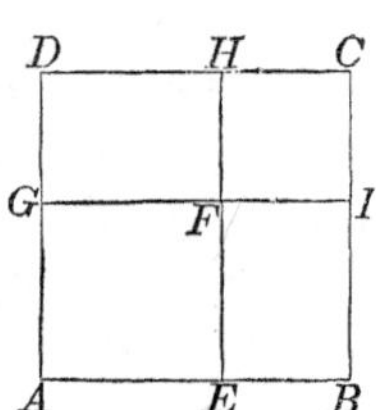

But the square $ABCD$ is made up of four parts, viz., the square on AE; the square on EB; the rectangle DF, and the rectangle FB. Hence, the square on AB is equivalent to the square on AE plus the square on EB, plus twice the rectangle contained by AE and EB.

Cor. If the line AB was divided into two equal parts, the rectangles DF and FB would become squares, and the square described on the whole line would be equivalent to four times the square described on half the line.

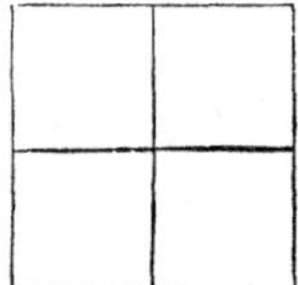

Sch. The property may be expressed in the language of algebra, thus,

$$(a+b)^2=a^2+2ab+b^2$$

THEOREM XII.

The square described on the hypothenuse of a right angled triangle, is equivalent to the sum of the squares described on the other two sides.

Let BAC be a right angled triangle, right angled at A: then will the square described on the hypothenuse BC, be equivalent to the two squares described on BA and AC.

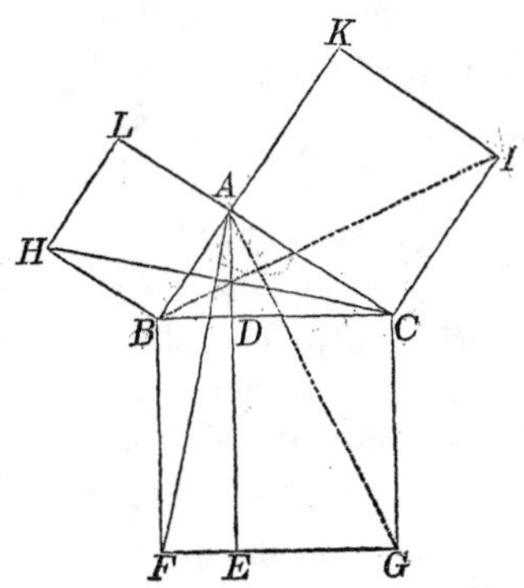

Having described the squares BG, BL and AI, let fall from A, on the hypothenuse, the perpendicular AD, and produce it to E; then draw the diagonals AF, CH.

Now, the angle ABF is made up of the right angle FBC and the angle CBA; and the angle CBH is made up of the right angle ABH and the same angle CBA: hence, the angle ABF is equal to CBH. But FB is equal to BC, being sides of the same square; and for a like reason, BA is equal to HB. Therefore, the two triangles ABF and CBH, having two sides and the included angle of the one equal to two sides and the included angle of the other, each to each, are equal (Bk. I. Th. iv).

Since the angles BAC and BAL are right angles, as also the angle ABH, it follows that CAL is a straight line parallel to BH. Hence, the square HA and the triangle HBC, stand on the same base and between the same parallels; therefore, the triangle is half the square (Th. iii. Cor.). For a like reason, the triangle ABF is half the rectangle BE.

But it has already been proved that the triangle ABF is equal to the triangle CBH: hence, the rectangle BE, which is double the former, is equivalent to the square BL, which is double the latter (Ax. 6).

In the same manner it may be proved, that the rectangle DG is equivalent to the square CK.

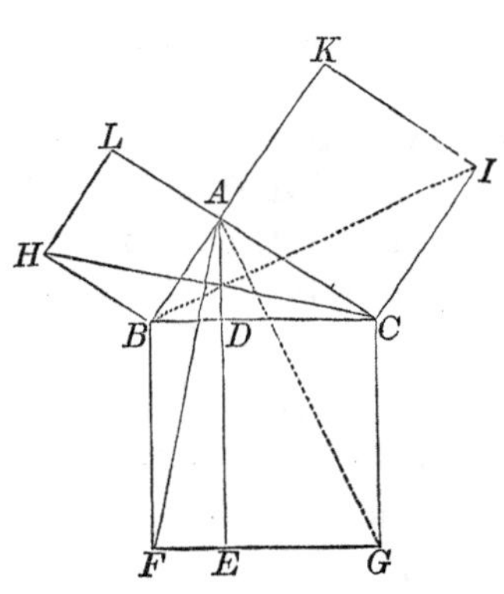

But the two rectangles BE, DG, make up the square BG: therefore, the square BG, described on the hypothenuse, is equivalent to the squares BL and CK, described on the other two sides.

Cor. Hence, the square of either side of a right angled triangle is equivalent to the square of the hypothenuse diminished by the square of the other side. That is, in the right angled triangle ABC

$$\overline{AB}^2=\overline{AC}^2-\overline{BC}^2$$

or

$$\overline{BC}^2=\overline{AC}^2-\overline{AB}^2.$$

Sch. The last theorem may be illustrated by describing a square on the hypothenuse BC, equal to 5, also on the sides BA, AC, respectively equal to 4 and 3; and observing that the number of small squares in the large square is equal to the number in the two small squares.

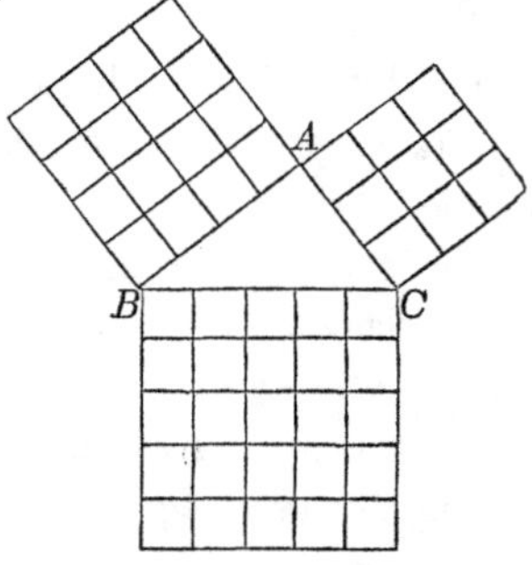

THEOREM XIII.

If a line be drawn parallel to the base of a triangle, it will divide the other two sides proportionally.

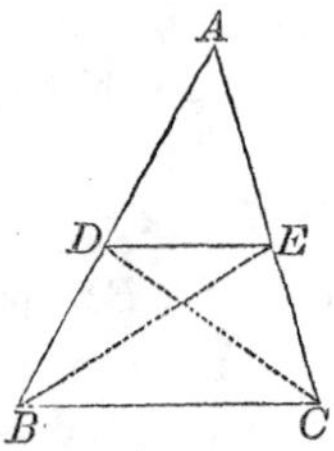

Let ABC be any triangle, and DE a straight line drawn parallel to the base BC: then will

$$AD : DB :: AE : EC.$$

For, draw BE and DC. Then, the two triangles BDE and DCE have the same base DE, and the same altitude, since their vertices B and C, lie in the line BC parallel to DE: hence, they are equivalent (Th. ii).

Again, the triangles ADE and BDE, having a common vertex E, have the same altitude; and consequently, are to each other as their bases (Th. ix, Cor.); hence, we have

$$ADE : BDE :: AD : DB.$$

But the triangles ADE and CDE, having a common vertex D, are to each other as their bases AE and EC: hence, we have

$$ADE : CDE :: AE : EC.$$

But the triangles BDE and CDE have been proved equivalent: hence, in the two proportions, the first antecedent and consequent in each are equal: therefore, by (Bk. III. Th. v), we have

$$AD : BD :: AE : EC.$$

Cor. The sides AB, AC, are also proportional to the parts AD, AE, or to BD, CE.

For, by composition (Bk. III. Th. vii), we have

$$AD+BD : BD :: AE+EC : EC.$$

Then, by alternation (Bk. III. Th. iv).

$AB : AC :: BD : EC$, hence, also, $AB : AC :: AD : AE$.

THEOREM XIV.

A line which bisects the vertical angle of a triangle divides the base into two segments which are proportional to the adjacent sides.

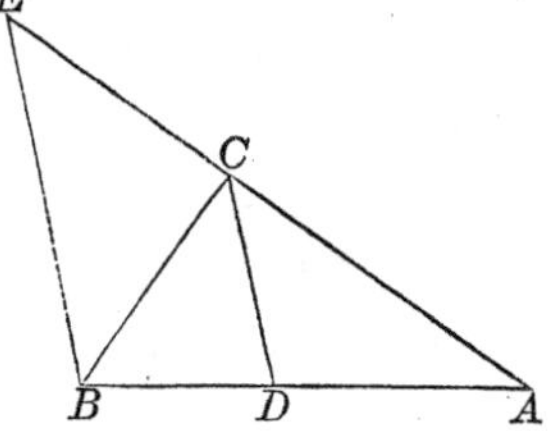

Let ACB be a triangle, having the angle C bisected by the line CD: then will

$AD : DB :: AC : CB.$

For, draw BE parallel to CD and produce AC to E. Then, since CB cuts the two parallels CD, EB, the alternate angles BCD and CBE are equal (Bk. I. Th. xii): hence, CBE is equal to angle ACD.

But, since AE cuts the two parallels CD, BE, the angle ACD is equal to CEB (Bk. I. Th. xiv): consequently, the angle CBE is equal to the angle CEB (Ax. 1): hence, the side CB is equal to CE (Bk. I. Th. vii).

Now, in the triangle ABE the line CD is drawn parallel to BE: hence, by the last theorem, we have

$$AD : DB :: AC : CE,$$

and by placing for CE, its equal CB, we have

$$AD : DB :: AC : CB.$$

THEOREM XV.

Equiangular triangles have their homologous sides proportional.

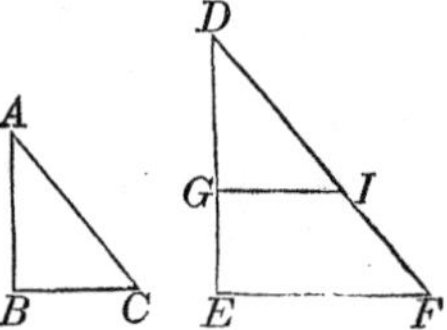

Let ABC and DEF be two equiangular triangles, having the angle A equal to the angle D, the angle C to the angle F, and the angle B to the angle E: then will

$AB : AC :: DE : DF.$

For, on the sides of the larger triangle DEF, make DI equal to AC and DG equal to AB, and join IG. Then the two triangles ABC and DIG, having two sides and the included angle of the one equal to two sides and the included angle of the other, each to each, will be equal (Bk. I Th. iv). Hence, the angles I and G are equal to C and B, and consequently, to the angles F and E: therefore, IG is parallel to EF (Bk. I. Th. xiv, Cor. 1).

Now, in the triangle DEF, since IG is parallel to the base, we have (Th. xiii).

$$DG : DI :: DE : DF,$$

that is,

$$AB : AC :: DE : DF.$$

THEOREM XVI.

Two triangles which have their homologous sides proportional are equiangular and similar.

Let BAC and EDF be two triangles having

$$BC : EF :: AB : ED,$$

and $BC : EF :: AC : DF$;

then will they have the homologous angles equal, viz., the angle

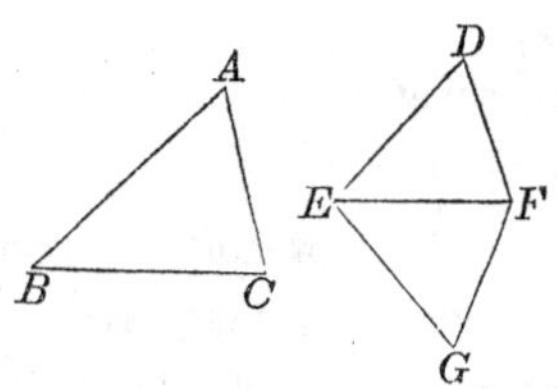

$$B{=}E, \quad A{=}D \quad \text{and} \quad C{=}F.$$

For, at the point E make FEG equal to the angle B; and at F make the angle EFG equal to the angle C: Then will the angle at G be equal to A, and the two triangles BAC and EGF will be equiangular (Bk. I. Th. xvii. Cor 1).

Therefore, by the last theorem we shall have

$$BC : EF :: AB : EG;$$

but by hypothesis,

$$BC : EF :: AB : DE:$$

hence, EG is equal to ED.

By the last theorem we also have

$$BC : EF :: AC : FG,$$

and by hypothesis,

$$BC : EF :: AC : DF;$$

hence, FG is equal to DF.

Therefore, the triangles DEF and EGF, having their three sides equal, each to each, are equiangular (Bk. I. Th. viii) But, by construction, the triangle EFG is equiangular with BAC: hence, the triangles BAC and EDF are equiangular, and consequently they are similar.

Sch. By Theorem XV, it appears that if the corresponding angles of two triangles are equal, each to each, the homologous sides will be proportional; and in the last theorem it was proved that if the sides are proportional, the corresponding angles will be equal.

Now, these proportions do not hold good in the quadrilaterals. For, in the square and rectangle, the corresponding angles are equal, but the sides are not proportional; and the angles of a parallelogram or quadrilateral, may be varied at pleasure, without altering the lengths of the sides.

THEOREM XVII.

If two triangles have an angle in the one equal to an angle in the other, and the sides containing these angles proportional, the two triangles will be equiangular and similar.

Let ABC and DEF be two triangles having the angle A equal to the angle D, and

$$AB : DE :: AC : DF;$$

then will the two triangles be similar.

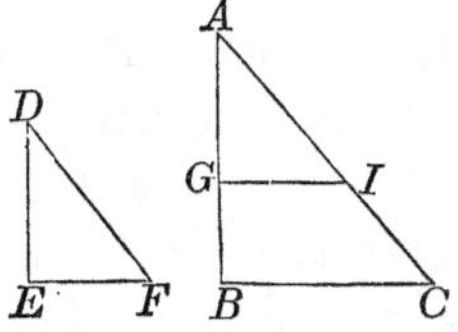

For, lay off AG equal to DE, and through G draw GI parallel to BC. Then the angle AGI will be equal to the angle ABC (Bk. I. Th. xiv); and the triangles AGI and ABC will be equiangular. Hence, we shall have

$$AB : AG :: AC : AI.$$

But, by hypothesis, we have

$$AB : DE :: AC : DF,$$

and by construction, AG is equal to DE; therefore, AI is equal to DF, and consequently, the two triangles AGI and DEF are equal in all their parts (Bk. I. Th. iv). But the triangle ABC is similar to AGI, consequently it is similar to DEF

THEOREM XVIII.

If from the right angle of a right angled triangle, a perpendicular be let fall on the hypothenuse, then

I. *The two partial triangles thus formed will be similar to each other and to the whole triangle.*

II. *Either side including the right angle will be a mean proportional between the hypothenuse and the adjacent segment.*

III. *The perpendicular will be a mean proportional between the segments of the hypothenuse.*

Let ABC be a right angled triangle, and AD perpendicular to the hypothenuse.

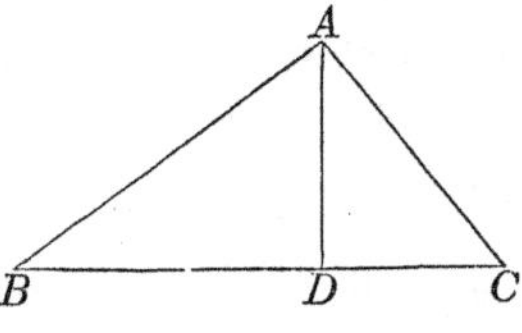

The two triangles BAC and BAD having the common angle B, and the right angle BAC equal to the right angle at D, will be equiangular (Bk. I. Th. xvii. Cor. 1); and, consequently, similar (Th. xv). For a like reason the triangles BAC and CAD are similar.

Now, from the triangles BAC and BAD, we have

$$BC : BA :: BA : BD.$$

From the triangles BAC and CAD, we have

$$BC : CA :: CA : CD;$$

and from the triangles BAD and DAC, we have

$$BD : AD :: AD : DC.$$

Cor. If from a point A, in the circumference of a circle, AD be drawn perpendicular to any diameter as BC, and the chords AB AC be also drawn, then the angle BAC will be a right angle (Bk. II. Th. x): and by the theorem we shall have,

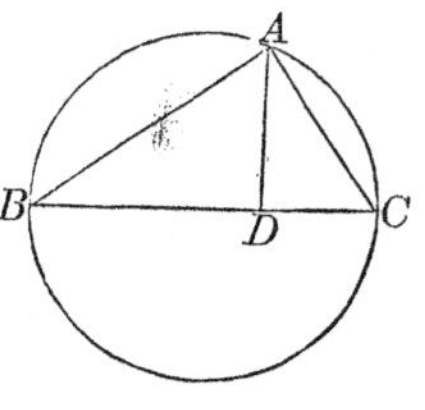

1st The perpendicular AD a mean proportional between the segments BD and DC.

2d Each chord will be a mean proportional between the diameter and the adjacent segment.

That is,

$$\overline{AD}^2 = BD \times DC$$

$$\overline{AB}^2 = BC \times BD$$

$$\overline{AC}^2 = BC \times CD.$$

THEOREM XIX.

Similar triangles are to each other as the squares described on their homologous sides.

Let ABC and DEF be two similar triangles, and AL and DN the squares described on the homologous sides AB, DE: then will the triangle

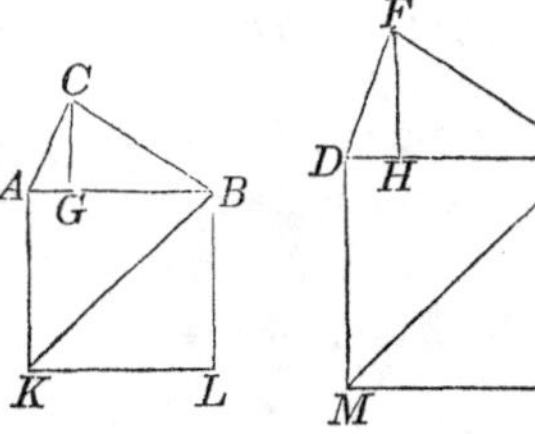

$ABC : DEF :: AL : DN.$

For, draw CG and FH perpendicular to the bases AB, DE, and draw the diagonals BK and EM.

Then, the similar triangles ABC and DEF, having their like sides proportional, we have

$$AC : DF :: AB : DE;$$

and the two ACG, DFH, give

$$AC : DF :: CG : FH;$$

hence, (Bk. III. Th. v), we have

$$AB : DE :: CG : FH,$$

or (Bk. III. Th. iv),

$$AB : CG :: DE : FH.$$

Now, the two triangles ABC and AKB have the common base AB; and the triangles DEF and DEM have the common base DE; and since triangles on equal bases are to each other as their altitudes (Th. ix, Cor.), we have

the triangle

$$ABC : ABK :: CG : AK \text{ or } AB$$

and the triangle,

$$DEF : DME :: FH : DM \text{ or } DE$$

But we have proved

$$CG : AB :: FH : DE;$$

hence, $ABC : ABK :: DEF : DME$,

or, alternately,

$$ABC : DEF :: ABK : DME.$$

But the squares AL and DN, being each double of the triangles AKB and DME will have the ratio ; hence,

$$ABC : DEF :: AL : DN.$$

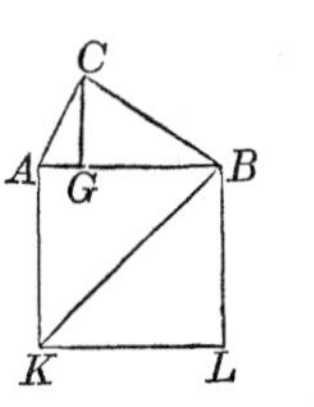

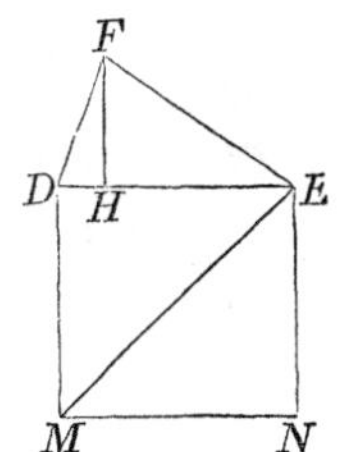

THEOREM XX.

Two similar polygons may be divided into an equal number of triangles, similar each to each, and similarly placed.

Let $ABCDE$ and $FGHIK$ be two similar polygons.

From the angle A draw the diagonals AC, AD: and from the homologous angle F, draw FH, FI.

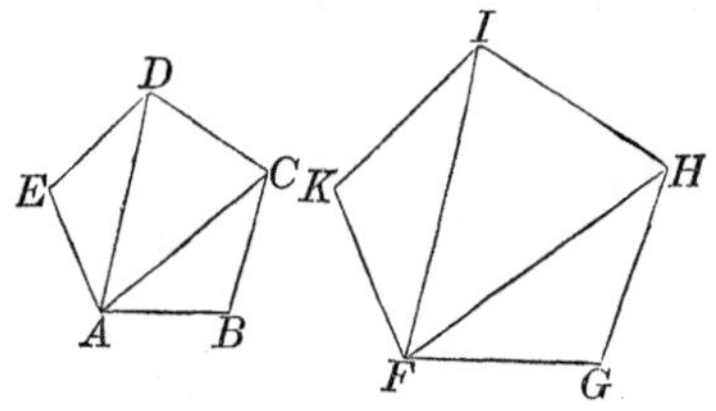

Now, since the polygons are similar, the homologous angles B and G will be equal, and the sides about the equal angles proportional (Def. 1): that is,

$$AB : BC :: FG : GH.$$

Hence, the triangles ABC and FGH have an angle in each equal, and the sides about the equal angles proportional: therefore, they are similar (Th. xvii), and consequently, the angle ACB is equal to FHG. Taking these from the equal angles BCD and GHI there will remain ACD equal to FHI. The two triangles ACD and FHI will then have an angle in each equal, and the sides about the equal angles proportional: hence, they will be similar.

In the same manner it may be shown that the triangles AED and FKI are similar: and, hence, whatever be the number of sides of the polygons, they may be divided into an equal number of similar triangles.

THEOREM XXI.

Similar polygons are to each other as the squares described on their homologous sides.

Let $ABCDE$ and $FGNIK$, be two similar polygons; then will they be to each other as the squares described on AB, FG, or any other two homologous sides.

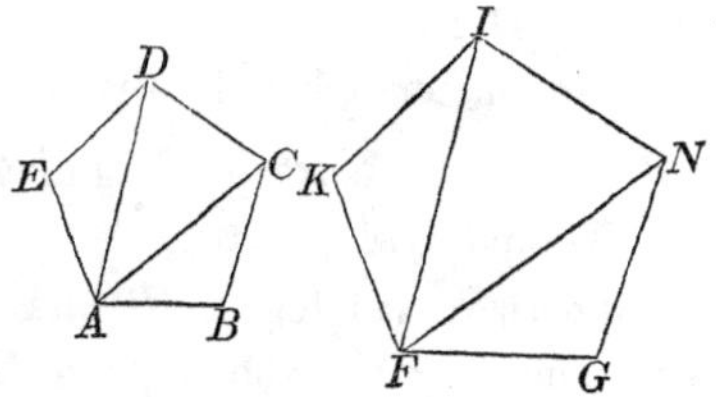

For, let the polygons be divided, as in the last theorem, into an equal number of similar triangles. Then, by Theorem XIX, we have the triangles

$$ABC : FGN :: \overline{AB}^2 : \overline{FG}^2$$

$$ADC : FIN :: \overline{DC}^2 : \overline{IN}^2$$

$$ADE : FIK :: \overline{DE}^2 : \overline{IK}^2$$

But since the polygons are similar, the ratio of the last antecedent to its consequent, in each of the proportions, is the same: hence, we have (Bk. III. Th. xii).

$$ABC+ADC+ADE : FGN+FIN+FIK :: \overline{AB}^2 : \overline{FG}^2;$$

that is, $ABCDE : FGNIK :: \overline{AB}^2 : \overline{FG}^2$;

Hence, the areas of similar polygons are to each other as the squares described on their homologous sides.

THEOREM XXII.

If similar polygons are inscribed in circles, their homologous sides, and also their perimeters, will have the same ratio to each other as the diameters of the circles in which they are inscribed.

Let $ABCDE$, $FGNIK$, be two similar figures, inscribed in the circles whose diameters are AL and FM: then will each side, AB, BC, &c., of the one, be to the homologous side FG, GN, &c., of the other, as the diameter AL to the diameter FM. Also, the perimeter $AB+BC+CD$ &c., will be to the perimeter $FG+GN+NI$ &c., as the diameter AL to the diameter FM.

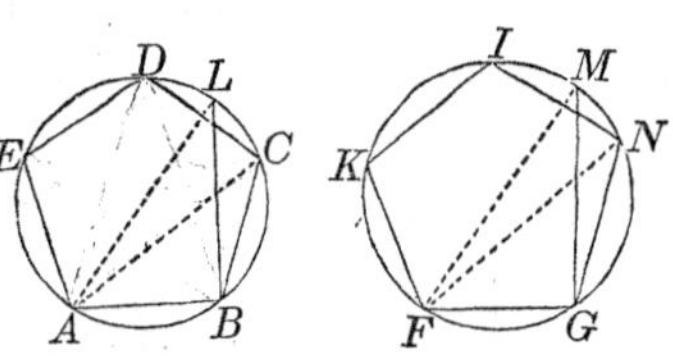

For, draw the two corresponding diagonals AC, FN, as also the lines BL and GM.

Then, the two triangles ACB and FNG will be similar (Th. xx); and therefore, the angle ACB is equal to the angle FNG. But, the angle ACB is equal to the angle ALB, and the angle FNG to the angle FMG (Bk. II. Th. ix): hence, the angle ALB is equal to the angle FMG (Ax. 1); and since ABL and FGM are right angles (Bk. II. Th. x), the two triangles ALB and FMG will be equiangular (Bk. I. Th. xvii. Cor. 1), and consequently similar (Th. xv).

Therefore,

$$AB : FG :: AL : FM.$$

Again, since any two homologous sides are to each other in the same ratio as AL to FM, we have (Bk. III. Th. xii),

$$AB+BC+CD \text{ \&c.} : FG+GN+NI \text{ \&c.} :: AL : FM$$

THEOREM XXIII.

Similar polygons inscribed in circles are to each other as the squares of the diameters of the circles.

Let $ABCDE$, $FGNIK$, be two polygons inscribed in the circles whose diameters are AL and FM: then will the polygon $ABCDE$, be to the polygon $FGNIK$ as the square of AL to the square of FM.

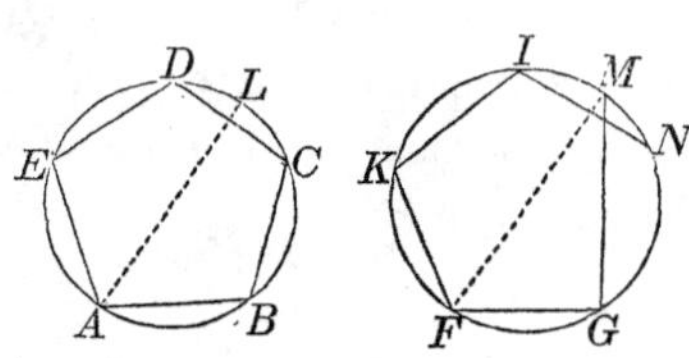

For, the polygons being similar, are to each other as the squares of their like sides (Th. xxi); that is, as $\overline{AB}^2$ to $\overline{FG}^2$.

But, by the last theorem,

$$AB : FG :: AL : FM;$$

therefore (Bk III. Th. xiii),

$$\overline{AB}^2 : \overline{FG}^2 :: \overline{AL}^2 : \overline{FM}^2;$$

consequently,

$$ABCDE : FGNIK :: \overline{AL}^2 : \overline{FM}^2.$$

Sch. If any regular polygon, $ABDEFG$, be inscribed in a circle, and then the arcs AB, BE, &c., be bisected, and lines be drawn through these points of bisection, a new polygon will be formed having double the number of sides. It is plain that this new polygon will differ less from the circle than the first polygon, and its sides will lie nearer the circumference than the sides of the first polygon.

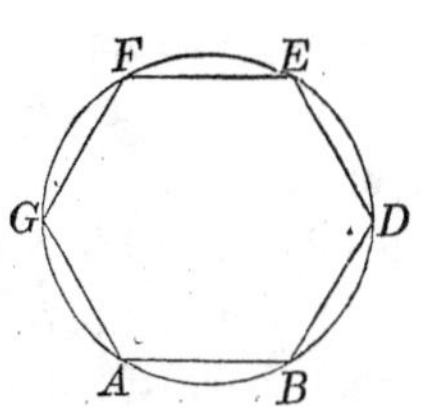

If now, we suppose the number of sides to be continually increased, the length of each side will constantly diminish, until finally the polygon will become equal to the circle, and the perimeter will coincide with the circumference. When this takes place, the line CH, drawn perpendicular to one of the sides, will become equal to the radius of the circle.

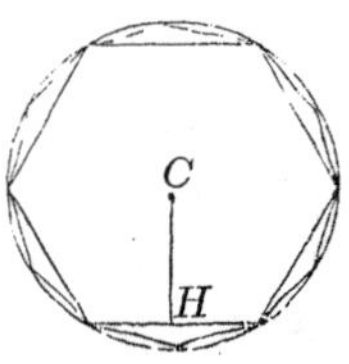

THEOREM XXIV.

The circumferences of circles are to each other as their diameters.

Let there be two circles whose diameters are AL and FM: then will their circumferences be to each other as AL to FM.

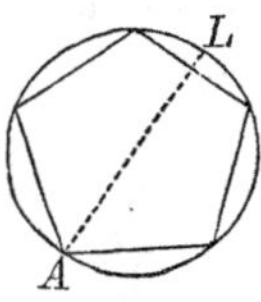

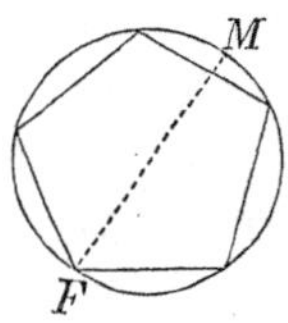

For, suppose two similar polygons to be inscribed in the circles: their perimeters will be to each other as AL to FM (Th. xxii).

Let us now suppose the arcs which subtend the sides of the polygons to be bisected, and new polygons of double the number of sides to be formed: their perimeters will still be to each other as AL to FM, and if the number of sides be increased until the perimeters coincide with the circumference, we shall have the circumferences to each other as the diameters AL and FM.

THEOREM XXV.

The areas of circles are to each other as the squares of their diameters.

Let there be two circles whose diameters are AL and FM: then will their areas be to each other as the square of AL to the square of FM.

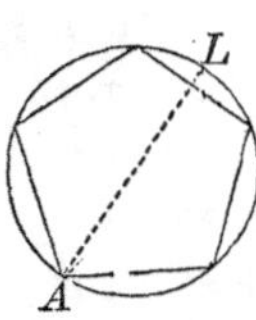

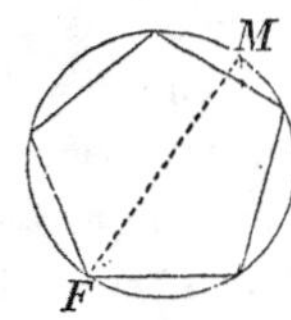

For, suppose two similar polygons to be inscribed in the circles: then will they be to each other as $\overline{AL}^2$ to $\overline{FM}^2$ (Th. xxiii).

Let us now suppose the number of sides of the polygons to be increased, by bisecting the arcs, until their perimeters shall coincide with the circumference of the circles. The polygons will then become equal to the circle, and hence, the areas of the circles will be to each other as the squares of their diameters.

Cor. Since the circumferences of circles are to each other as their diameters (Th. xxiv), it follows, that the areas which are proportional to the squares of the diameters, will also be proportional to the squares of the circumferences.

THEOREM XXVI.

The area of a regular polygon inscribed in a circle, is equal to half the product of the perimeter and the perpendicular let fall from the centre on one of the sides.

Let C be the centre of a circle circumscribing the regular polygon, and CD a perpendicular to one of its sides: then will its area be equal to half the product of CD by the perimeter.

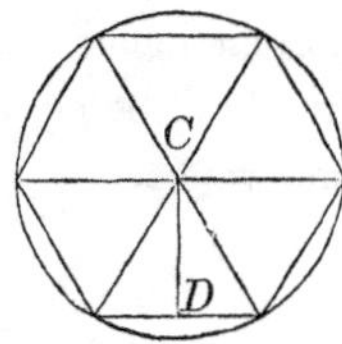

For, from C draw radii to the vertices of the angles, forming as many equal triangles as the polygon has sides, in each of which the perpendicular on the base will be equal to CD. Now, the area of one of them, as ACB, will be equal to half the product of CD by the base AB; and the same will be true for each of the other triangles: hence, the area of the polygon will be equal to half the product of CD by the perimeter.

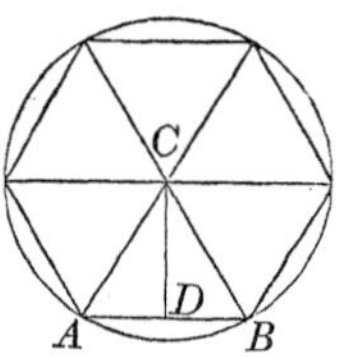

THEOREM XXVII.

The area of a circle is equal to half the product of the radius by the circumference.

Let C be the centre of a circle: then will its area be equal to half the product of the radius AC by the circumference ABE.

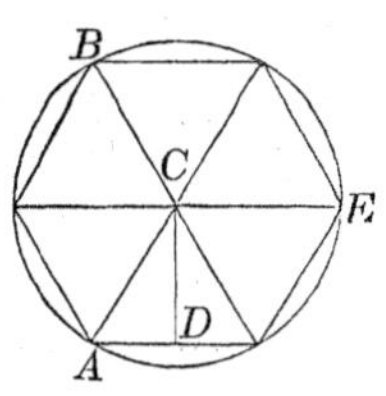

For, inscribe within the circle a regular hexagon, and draw CD perpendicular to one of its sides. Then, the area of the polygon will be equal to half the product of CD multiplied by the perimeter (Th. xxvi).

Let us now suppose the number of sides of the polygons to be increased, until the perimeter shall coincide with the circumference; the polygon will then become equal to the circle, and the perpendicular CD to the radius CA. Hence, the area of the circle will be equal to half the product of the radius by the circumference.

PROBLEMS

RELATING TO THE FOURTH BOOK.

PROBLEM I.

To divide a line into any proposed number of equal parts.

Let AB be the line, and let it be required to divide it into four equal parts.

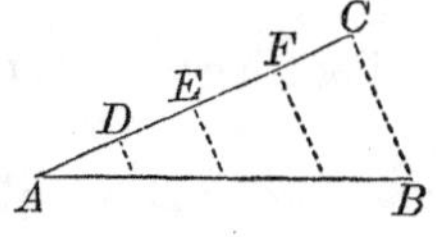

Draw any other line, AC, forming an angle with AB, and take any distance, as AD, and lay it off four times on AC. Join C and B, and through the points D, E, and F, draw parallels to CB. These parallels to BC will divide the line AB into parts proportional to the divisions on AC (Th. xiii): that is, into equal parts.

PROBLEM II.

To find a third proportional to two given lines.

Let A and B be the given lines.

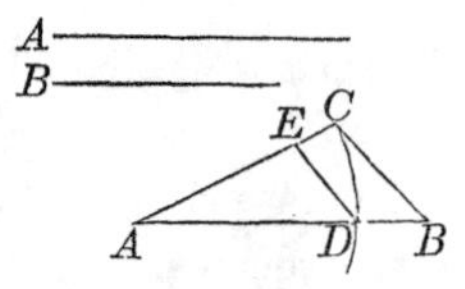

Make AB equal to A, and draw AC, making an angle with it. On AC lay off AC equal to B, and join BC: then lay off AD, also equal to B, and through D draw DE parallel to BC: then will AE be the third proportional sought.

For, since DE is parallel to BC, we have (Th. xiii).

$$AB \ : \ AC \ :: \ AD \ \text{or} \ AC \ : \ AE;$$

therefore, AE is the third proportional sought.

PROBLEM III.

To find a fourth proportional to the lines A, B, and C.

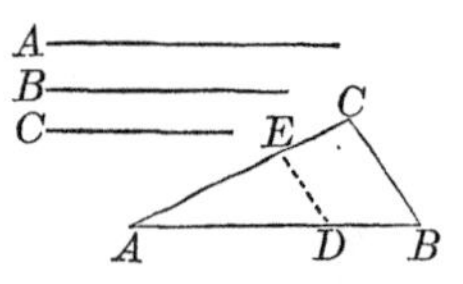

Place two of the lines forming an ngle with each other at A; that is, make AB equal to A, and AC equal B; also, lay off AD equal to C. Then join BC, and through D draw DE parallel to BC, and AE will be the fourth proportional sought.

For, since DE is parallel to BC, we have

$$AB : AC :: AD : AE;$$

therefore, AE is the fourth proportional sought.

PROBLEM IV.

To find a mean proportional between two given lines, A and B.

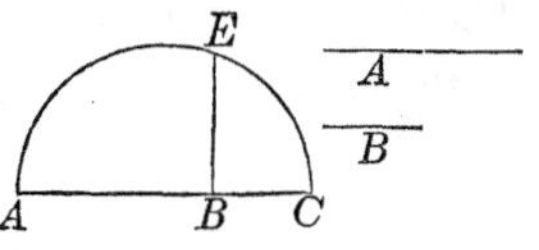

Make AB equal to A, and BC equal to B: on AC describe a semicircle. Through B draw BE perpendicular to AC, and it will be the mean proportional sought (Th. xviii. Cor).

PROBLEM V.

To make a square which shall be equivalent to the sum of two given squares.

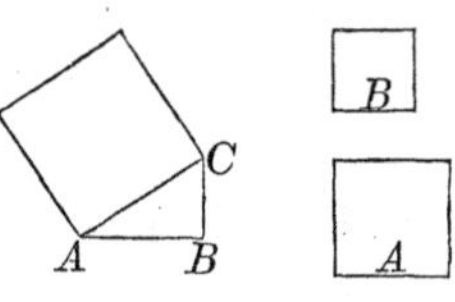

Let A and B be the sides of the given squares.

Draw an indefinite line AB, and make AB equal to A. At B draw BC perpendicular to AB, and make BC equal to B: then draw AC, and the square described on AC will be equivalent to the squares on A and B (Th. xii).

PROBLEM VI.

To make a square which shall be equivalent to the difference between two given squares.

Let A and B be the sides of the given squares.

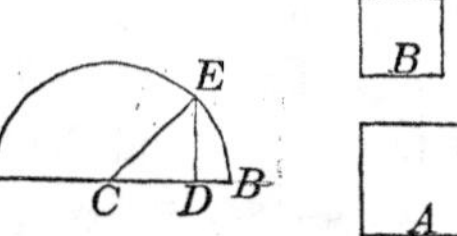

Draw an indefinite line, and make CB equal to A, and CD equal to B. At D draw DE perpendicular to CB, and with C as a centre, and CB as a radius, describe a semicircle meeting DE in E, and join CE: then will the square described on ED be equal to the difference between the given squares.

For, CE is equal to CB, that is, equal to A, and CD is equal to B: and by (Th. xii. Cor.),

$$\overline{ED}^2 = \overline{CE}^2 - \overline{CD}^2.$$

PROBLEM VII.

To make a triangle which shall be equivalent to a given quadrilateral.

Let $ABCD$ be the given quadrilateral.

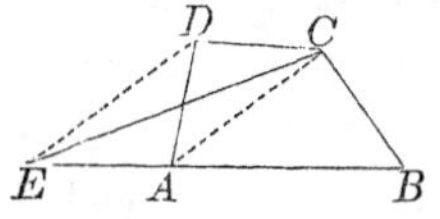

Draw the diagonal AC, and through D draw DE parallel to AC, meeting BA produced at E. Join EC: then will the triangle CEB be equivalent to the quadrilateral BD.

For, the two triangles ACE and ADC, having the same base AC, and the vertices of the angles D and E in the same line DE parallel to AC, are equivalent (Th. ii). If to each, we add ACB, we shall then have the triangle ECB equal to the quadrilateral BD (Ax. 2).

PROBLEM VIII.

To make a triangle which shall be equivalent to a given polygon.

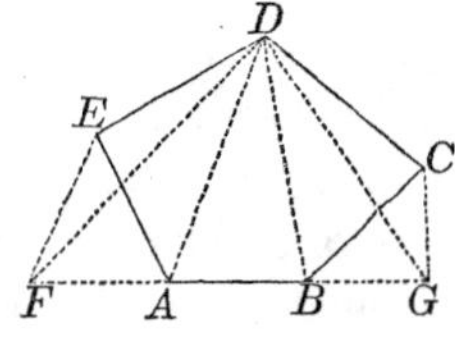

Let $ABCDE$ be the polygon.

Draw the diagonals AD, BD. Produce AB in both directions, and through C and E draw CG and EF, respectively parallel to AD and BD: then join FD and DG, and the triangle FDG will be equivalent to the polygon $ABCDE$.

For, the triangle ADE is equivalent to the triangle DAF, and DBC to DBG (Th. ii); and by adding ADB to the equals, we shall have the triangle FDG equivalent to the polygon $ABCDE$.

PROBLEM IX.

To make a rectangle that shall be equivalent to a given triangle

Let ABC be the given triangle.

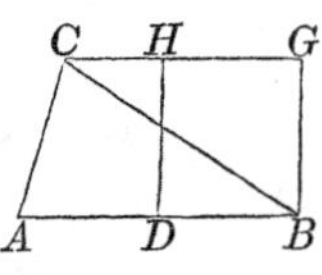

Bisect the base AB at D, and draw DH perpendicular to AB. Through C, the vertex of the triangle, draw CHG parallel to AB, and draw BG perpendicular to it: then will the rectangle DG be equivalent to the triangle ABC.

For, the triangle would be half a rectangle having the same base and altitude: hence, it is equivalent to DG, having half the same base and the same altitude.

PROBLEM X.

To inscribe a circle in a regular polygon.

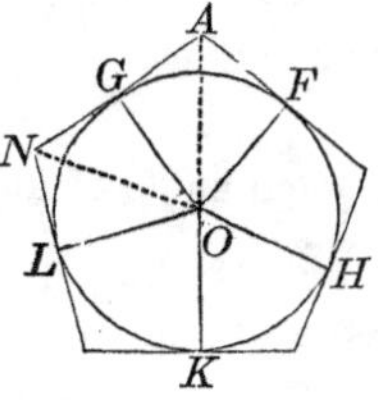

Bisect any two sides of the polygon by the perpendiculars GO, FO, and with their point of intersection O, as a centre, and OG as a radius describe the circumference of a circle—this circle will touch all the sides of the polygon.

For, draw OA. Then in the two right angled triangles OAG and OAF, the side AO is common, and AG is equal to AF, since each is half of one of the equal sides of the polygon: hence, OG is equal to OF (Bk. I. Th. xix). In the same manner it may be shown that OH, OK and OL are all equal to each other: hence, a circle described with the centre O and radius OF will be inscribed in the polygon.

Cor. Hence, also the lines OA, ON &c., drawn to the angles of the polygon are equal.

APPENDIX

OF THE REGULAR POLYGONS.

1. In a regular polygon the angles are all equal to each other (Def. 3). If then, the sum of the inward angles of a regular polygon be divided by the number of angles, the quotient will be the value of one of the angles.

But the sum of the inward angles is equal to twice as many right angles, wanting four, as the polygon has sides, and we shall find the value in degrees by simply placing 90° for the right angle.

2. Thus, for the sum of all the angles of an equilateral triangle, we have

$$6\times 90^\circ-4\times 90^\circ=540^\circ-360^\circ=180^\circ$$

and for each angle

$$180^\circ\div 3=60^\circ:$$

Hence, each angle of an equilateral triangle, is equal to 60 degrees.

3. For the sum of all the angles of a square, we have

$$8\times 90^\circ-4\times 90^\circ=720^\circ-360^\circ=360^\circ,$$

and for each of the angles

$$360^\circ\div 4=90^\circ$$

4. For the sum of all the angles of a regular pentagon, we have

$$10\times 90^\circ-4\times 90^\circ=900^\circ-360^\circ=540^\circ,$$

and for each angle

$$540^\circ\div 5=108^\circ.$$

5. For the sum of all the angles of a regular hexagon, we have

$$12\times 90^\circ-4\times 90^\circ=1080^\circ-360^\circ=720^\circ,$$

and of each angle

$$720^\circ\div 6=120^\circ.$$

6. For the sum of the angles of a regular heptagon, we have

$$14\times 90^\circ-4\times 90^\circ=1260^\circ-360^\circ=900^\circ:$$

and for one of the angles

$$900^\circ\div 7=128^\circ\ 34'+.$$

7. For the sum of the angles of a regular octagon, we have

$$16\times 90^\circ-4\times 90^\circ=1440^\circ-360^\circ=1080^\circ:$$

and for each angle

$$1080^\circ\div 8=135^\circ.$$

8. Since the sum of the angles about any point is equal to four right angles (Bk. I. Th. ii. Cor. 3), it may be observed tha there are only three kinds of regular polygons, which can be arranged around any point, as C, so as exactly to fill up the space. These are,

First.—Six equilateral triangles, in which each angle about C is equal to 60°, and their sum to

$$60° \times 6 = 360.$$

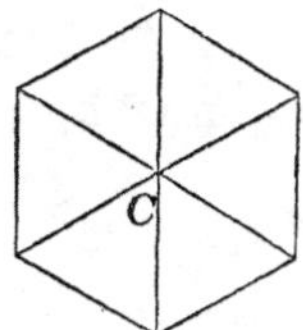

Second.—Four squares, in which each angle is equal to 90°, and their sum to

$$90° \times 4 = 360°$$

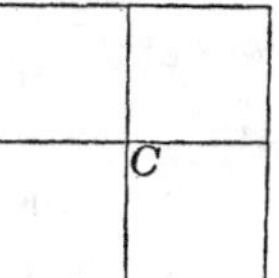

Third.—Three hexagons, in which each angle is equal to 120, and the sum of the three to

$$120° \times 3 = 360°.$$

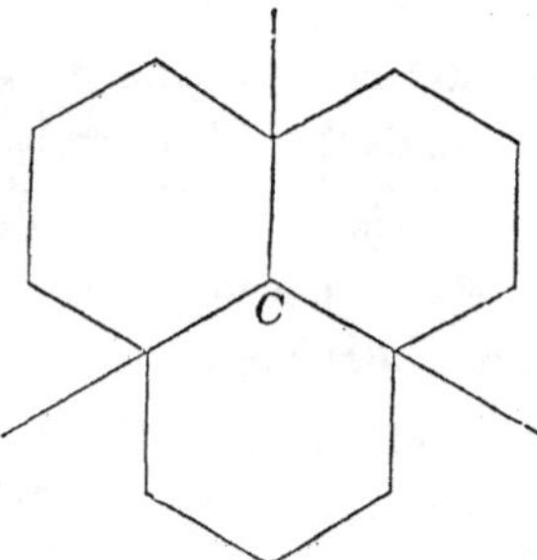

GEOMETRY.

BOOK V.

OF PLANES AND THEIR ANGLES.

DEFINITIONS.

1. A straight line is *perpendicular to a plane*, when it is perpendicular to every straight line of the plane which it meets. The point at which the perpendicular meets the plane, is called the *foot* of the perpendicular.

2. If a straight line is perpendicular to a plane, the plane is also said to be perpendicular to the line.

3. A line is parallel to a plane when it will not meet that plane, to whatever distance both may be produced. Conversely, the plane is then parallel to the line.

4. Two planes are parallel to each other, when they will not meet, to whatever distance both are produced.

5. If two planes are not parallel, they intersect each other in a line that is common to both planes: such line is called their *common intersection*.

6. The angle, or *inclination* of two planes, is measured by wo lines, one in each plane, and both perpendicular to the common intersection at the same point.

This angle may be acute, obtuse, or a right angle. When it is a right angle, the planes are said to be perpendicular to each other.

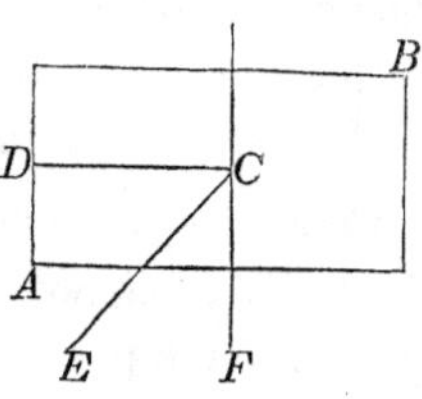

Let AB be a plane coinciding with the plane of the paper, and ECF a plane intersecting it in the line FC. Now, if from any point of the common intersection as C, we draw CD in the plane AB, and CE in the plane ECF, and both perpendicular to CF at C, then will the angle DCE measure the inclination between the two planes.

It should be remembered that the line EC is directly over the line CD.

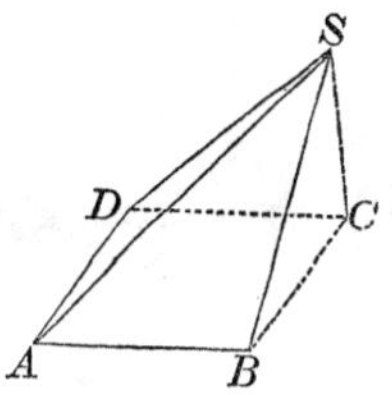

7. A solid angle is the angular space included between several planes meeting at the same point.

Thus, the solid angle S is formed by the meeting of the planes ASB, BSC, CSD, DSA.

Three planes, at least, are requisite to form a solid angle.

THEOREM I.

Two straight lines which intersect each other, lie in the same plane, and determine its position.

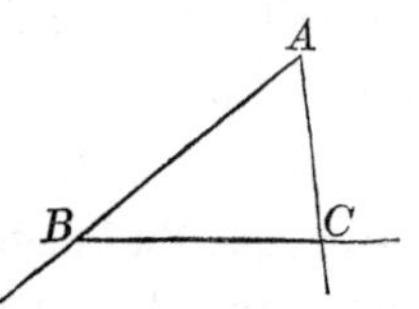

Let AB and AC be two straight lines which intersect each other at A.

Through AB conceive a plane to be passed, and let this plane be turned around AB until it embraces the point C: the plane will then contain the two lines AB, AC, and if it be turned either way it will depart from the point C, and consequently from the line AC. Hence, the position of the plane is determined by the single condition of containing the two straight lines AB, AC.

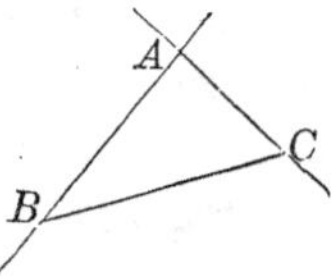

Cor. 1. A triangle ABC, or three points A, B, C, not in a straight line, determine the position of a plane.

Cor. 2. Hence, also, two parallels AB, CD determine the position of a plane. For drawing EF, we see that the plane of the two straight lines AE, EF is that of the parallels AB, CD.

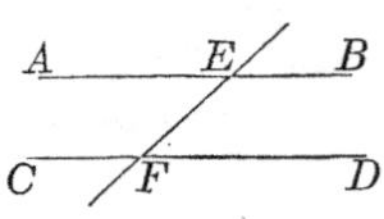

THEOREM II.

A perpendicular is the shortest line which can be drawn from a point to a plane.

Let A be a point above the plane DE, and AB a line drawn perpendicular to the plane: then will AB be shorter than any oblique line AC.

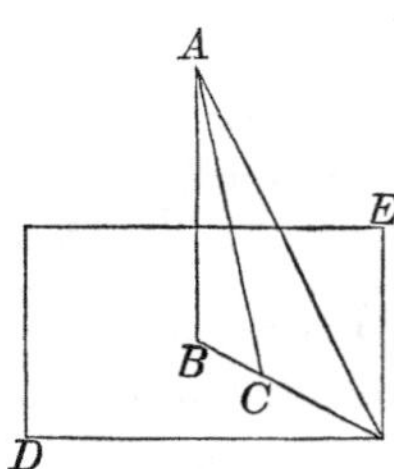

For, through B, the foot of the perpendicular, draw BC to the point where the oblique line AC meets the plane.

Now, since AB is perpendicular to the plane, the angle ABC will be a right angle (Def. 1.), and consequently less than the angle C: therefore, AB, opposite the angle C, will be less than AC, opposite the angle B (Bk. I. Th. xi).

Cor. It is evident that if several lines be drawn from the point A to the plane, that those which are nearest the perpendicular AB, will be less than those more remote.

Sch. The distance from a point to a plane is measured on the perpendicular: hence, when the *distance* only is named, the shortest distance is always understood.

THEOREM III.

The common intersection of two planes is a straight line.

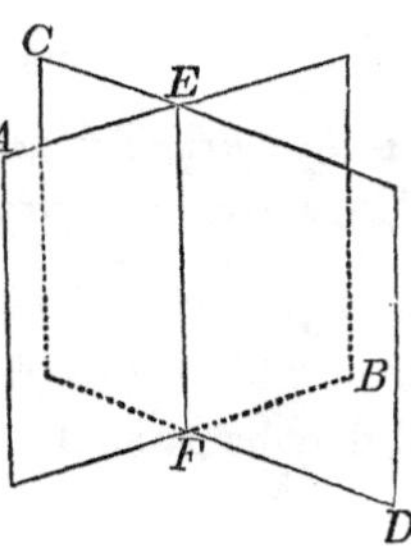

Let the two planes AB, CD, cut each other. Join any two points E and F, in the common intersection, by the straight line EF. This line will lie wholly in the plane AB, and also wholly in the plane CD (Bk. I. Def. 7); therefore, it will be in both planes at once, and consequently, is their common intersection.

THEOREM IV.

A straight line which is perpendicular to two straight lines at their point of intersection, will be perpendicular to the plane of those lines.

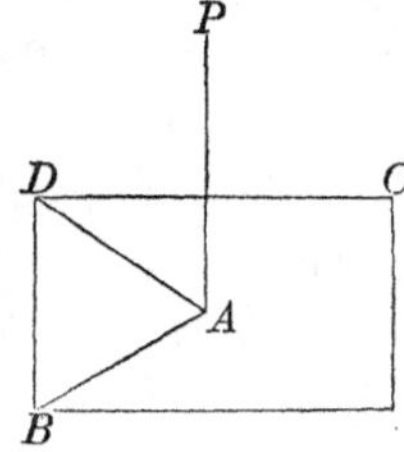

Let the line PA be perpendicular to the two lines AD, AB: then will it be perpendicular to the plane BC which contains them.

For, if AP is not perpendicular to the plane BC, suppose a plane to be drawn through A, that shall be perpendicular to AP.

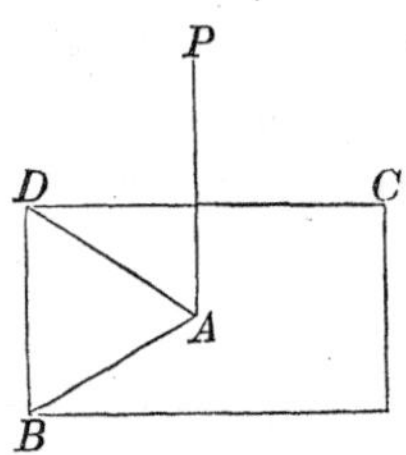

Now, every line drawn through A, and perpendicular to AP, will be a line of this last plane (Def. 1): hence, this last plane will contain the lines AB, AD, and consequently, a line which is perpendicular to two lines at the point of intersection, will be perpendicular to the plane of those lines.

THEOREM V.

If two straight lines are perpendicular to the same plane they will be parallel to each other.

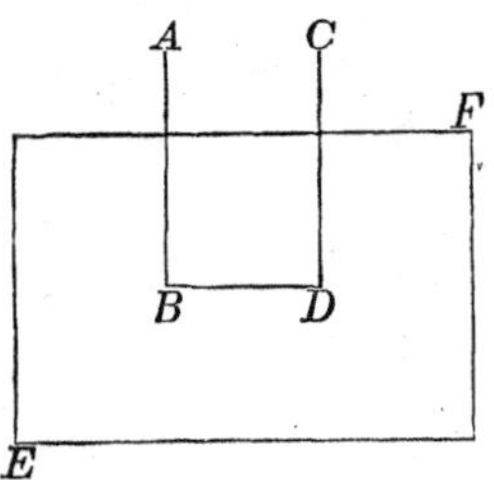

Let the two lines AB, CD, be perpendicular to the plane EF: then will they be parallel to each other.

For, join the points B and D, in which the lines meet the plane EF.

Then, because the lines AB, CD, are perpendicular to the plane EF, they will be perpendicular to the line BD (Def. 1); and since they are both contained in the plane $ABDC$ (Th. ii. Cor. 2), they will be parallel to each other (Bk. I. Th. xiii Cor.)

Cor. If two lines are parallel, and one of them is perpendicular to a plane, the other will also be perpendicular to the same plane.

THEOREM VI.

If two planes intersect each other at right angles, and a line be drawn in one plane perpendicular to the common intersection, this line will be perpendicular to the other plane.

Let the plane *FE* be perpendicular to *MN*, and *AP* be drawn in the plane *FE*, and perpendicular to the common intersection *DE:* then will *AP* be perpendicular to the plane *MN*.

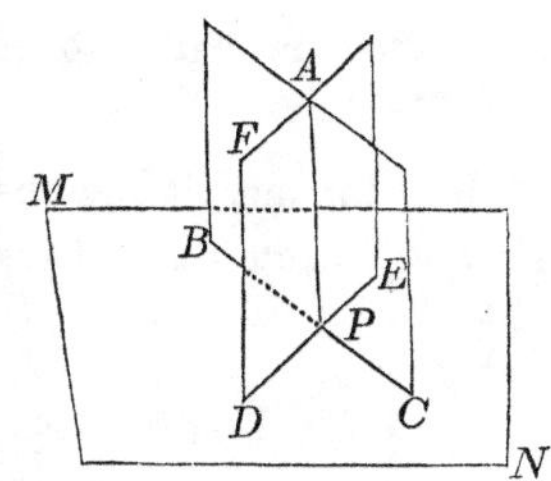

For, in the plane *MN* draw *CP* perpendicular to the common intersection *DE*. Then, because the planes *MN* and *FE* are perpendicular to each other, the angle *APC*, which measures their inclination, will be a right angle (Def. 6). Therefore, the line *AP* is perpendicular to the two straight lines *PC* and *PD;* hence, it is perpendicular to their plane *MN* (Th. iv)

THEOREM VII.

If one plane intersects another plane, the sum of the angles on the same side will be equal to two right angles.

Let the plane *GEF* intersect the plane *AB* in the line *FE:* then will the sum of the two angles on the samė side be equal to two right angles.

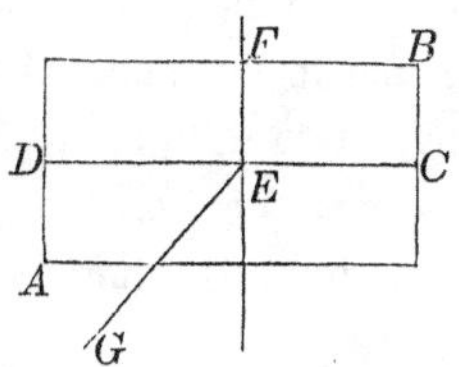

For, from any point, as *E*, in the common intersection, draw the lines *EG* and *DEC*, one in each plane, and both perpendicular to the common intersection at *E*. Then, the line *GE* makes, with the line *DEC*, two angle which together are equal to two right angles (Bk I. Th. ii): but these angles measure the inclination of the planes; therefore, the sum of the angles on the same side, which two planes make with each other, is equal to two right angles.

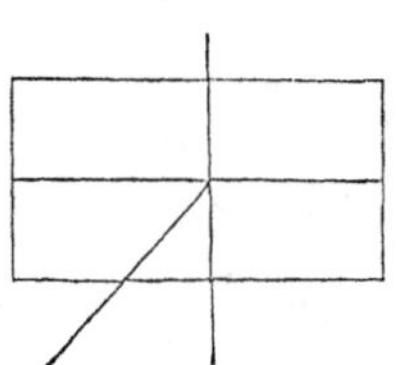

Cor. In like manner it may be demonstrated, that planes which intersect each other have their vertical or opposite angles equal.

THEOREM VIII.

Two planes which are perpendicular to the same straight line are parallel to each other.

Let the planes *MN* and *PQ* be perpendicular to the line *AB*: then will they be parallel.

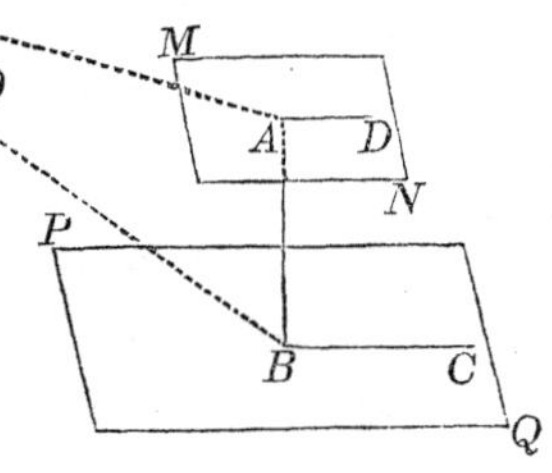

For, if they can meet any where, let *O* be one of their their common points, and draw *OB*, in the plane *PQ*, and *OA*, in the plane *MN*.

Now, since *AB* is perpendicular to both planes, it will be perpendicular to *OB* and *OA* (Def. 1): hence, the triangle *OAB* will have two right angles, which is impossible (Bk. I. Th. xvii. Cor. 4); therefore, the planes can have no point, as O, in common, and consequently, they are parallel (Def. 4).

THEOREM IX.

If a plane cuts two parallel planes, the lines of intersection will be parallel.

Let the parallel planes *MN* and *PA* be intersected by the plane *EH:* then will the lines of intersection *EF*, *GH*, be parallel.

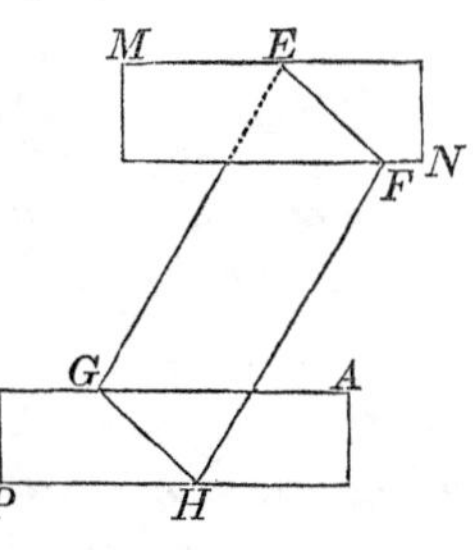

For, if the lines *EF*, *GH*, were not parallel, they would meet each other if sufficiently produced, since they lie in the same plane. If this were so, the planes *MN*, *PA*, would meet each other, and, consequently, could not be parallel; which would be contrary to the supposition.

THEOREM X.

If two lines are parallel to a third line, though not in the same plane with it, they will be parallel to each other.

Let the lines *AB* and *CD* be each parallel to the third line *EF*, though not in the same plane with it: then will they be parallel to each other.

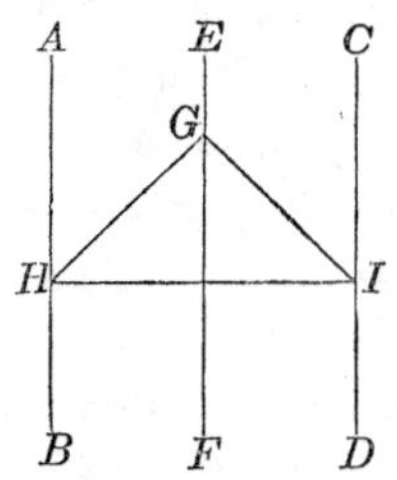

For, since *EF* and *CD* are parallel, they will lie in the same plane *FC* (Th. i. Cor. 2), and *AB*, *EF* will also lie in the plane *EB*.

At any point, *G*, in the line *EF*, let *GI* and *GH* be drawn in the planes *FC*, *BE*, and each perpendicular to *FE* at *G*.

Then, since the line *EF* is perpendicular to the lines *GH*, *GI*, it will be perpendicular to the plane *HGI* (Th. iv). And since *FE* is perpendicular to the plane *HGI*, its parallels *AB* and *DC* will also be perpendicular to the same plane (Th. v). Hence, since the two lines *AB*, *CD*, are both perpendicular to the plane *HGI*, they will be parallel to each other.

THEOREM XI.

If two angles, not situated in the same plane, have their sides parallel and lying in the same direction, the angles will be equal.

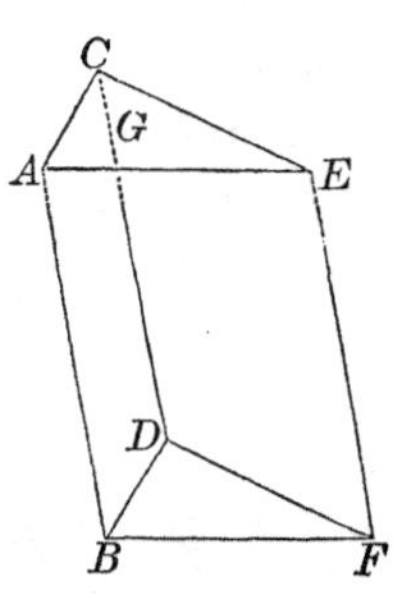

Let the angles ACE and BDF have the sides AC parallel to BD, and CE to DF: then will the angle ACE be equal to the angle BDF.

For, make AC equal to BD, and CE equal to DF, and join AB, CD, and EF; also, draw AE, BF.

Now since AC is equal and parallel to BD, the figure AD will be a parallelogram (Bk. I. Th. xxv); therefore, AB is equal and parallel to CD.

Again, since CE is equal and parallel to DF, CF will be a parallelogram, and EF will be equal and parallel to CD. Then, since AB and EF are both parallel to CD, they will be parallel to each other (Th. x); and since they are each equal to CD, they will be equal to each other. Hence, the figure $BAEF$ is a parallelogram (Bk. I. Th. xxv), and consequently, AE is equal to BF. Hence, the two triangles ACE and BDF have the three sides of the one equal to the three sides of the other, each to each, and therefore the angle ACE is equal to the angle BDF (Bk. I. Th. viii).

THEOREM XII.

If two planes are parallel, a straight line which is perpendicular to the one will also be perpendicular to the other.

Let MN and PQ be two parallel planes, and let AB be perpendicular to MN: then will it be perpendicular to PQ.

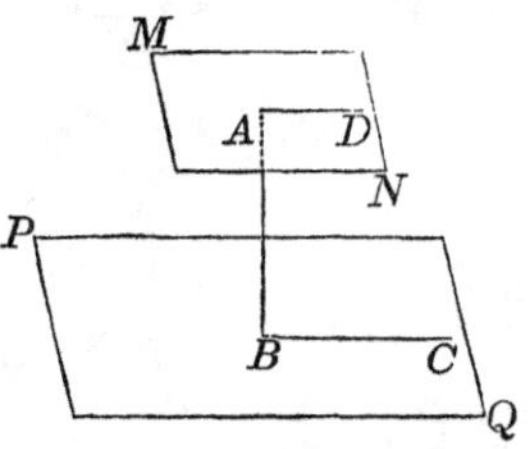

For, draw any line, BC, in the plane PQ, and through the lines AB, BC, suppose the plane ABC to be drawn, intersecting the plane MN in the line AD: then, the intersection AD will be parallel to BC (Th. ix). But since AB is perpendicular to the plane NM, it will be perpendicular to the straight line AD, and consequently, to its parallel BC (Bk. I. Th. xii. Cor.)

In like manner, AB might be proved perpendicular to any other line of the plane PQ, which should pass through B; hence, it is perpendicular to the plane (Def. 1).

GEOMETRY.

BOOK VI.

OF SOLIDS.

DEFINITIONS.

1. Every solid bounded by planes is called a *polyedron.*

2. The planes which bound a polyedron are called *faces.* The straight lines in which the faces intersect each other, are called the *edges* of the polyedron, and the points at which the edges intersect, are called the *vertices* of the angles, or vertices of the polyedron.

3. Two polyedrons are similar, when they are contained by the same number of similar planes, similarly situated, and equally inclined to each other.

4. A prism is a solid, whose ends are equal polygons, and whose side faces are parallelograms.

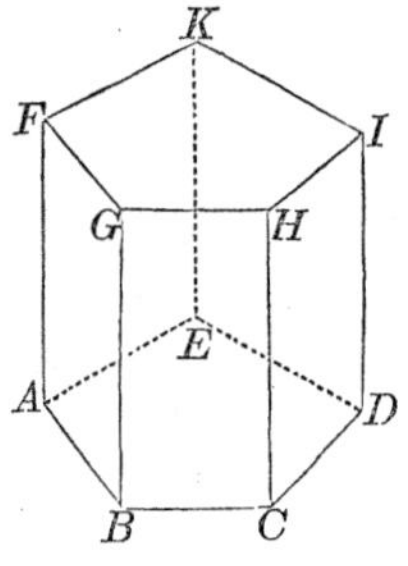

Thus, the prism whose lower base is the pentagon $ABCDE$, terminates in an equal and parallel pentagon $FGHIK$, which is called the *upper base.* The side faces of the prism are the parallelograms DH, DK, EF, AG, and BH. These are called the *convex*, or *lateral* surface of the prism.

5. The altitude of a prism is the distance between its upper and lower bases: that is, it is a line drawn from a point of the upper base, perpendicular, to the lower base.

6, A right prism is one in which the edges AF, BG, EK, HC, and DI, are perpendicular to the bases. In the right prism, either of the perpendicular edges is equal to the altitude. In the oblique prism the altitude is less than the edge.

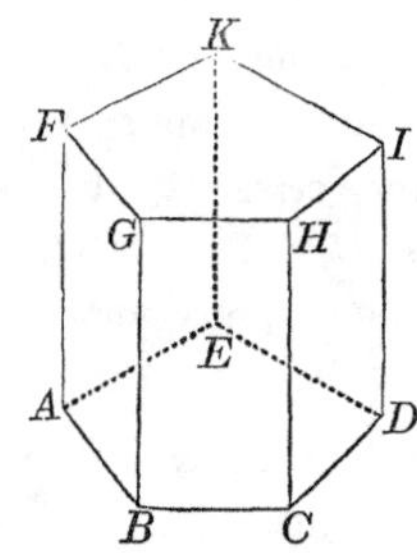

7. A prism whose base is a triangle, is called a *triangular* prism; if the base is a quadrangle, it is called a quadrangular prism; if a pentagon, a pentagonal prism; if a hexagon a hexagonal prism; &c.

8. A prism whose base is a parallelogram, and all of whose faces are also parallelograms, is called a parallelopipedon. If all the faces are rectangles, it is called a rectangular parallelopipedon

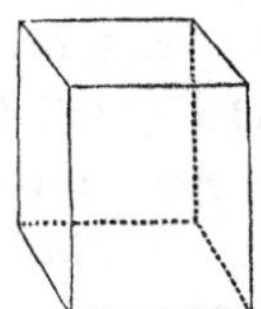

9. If the faces of the rectangular parallelopipedon are squares, the solid is called a *cube:* hence, the cube is a prism bounded by six equal squares

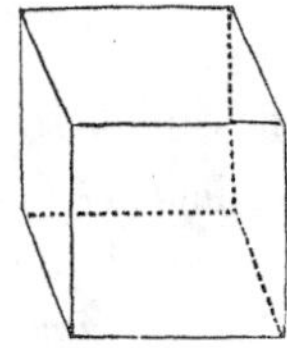

10. A pyramid is a solid, formed by several triangles united at the same point S, and terminating in the different sides of a polygon $ABCDE$.

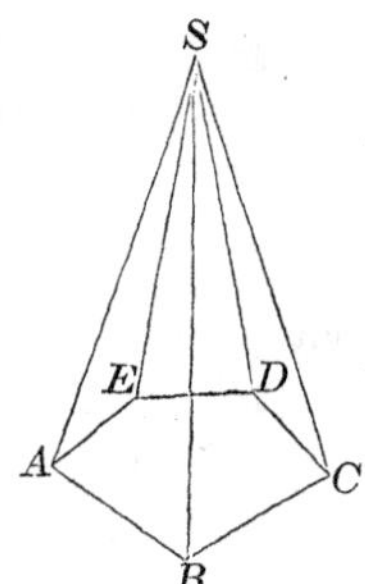

The polygon $ABCDE$, is called the *base* of the pyramid; the point S, is called the *vertex*, and the triangles ASB, BSC, CSD, DSE, and ESA, form its *lateral*, or *convex* surface.

11. A pyramid whose base is a triangle, is called a *triangular* pyramid; if the base is a quadrangle, it is called a quadrangular pyramid; if a pentagon, it is called a petagonal pyramid; if the base is a hexagon, it is called a hexagonal pyramid; &c.

12. The *altitude* of a pyramid, is the perpendicular let fall from the vertex, upon the plane of the base. Thus, SO is the altitude of the pyramid $S—ABCDE$.

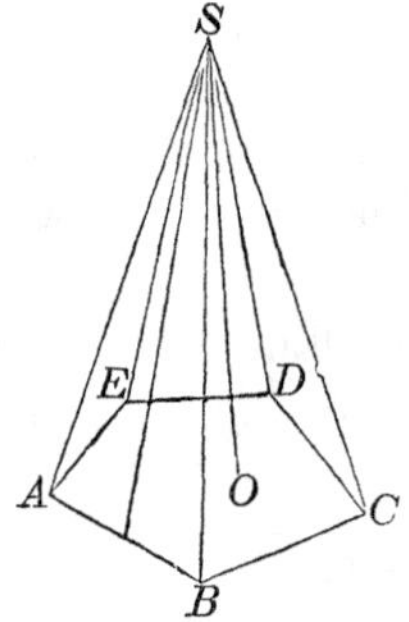

13. When the base of a pyramid is a regular polygon, and the perpendicular SO passes through the middle point of the base, the pyramid is called a regular pyramid, and the line SO is called the *axis*.

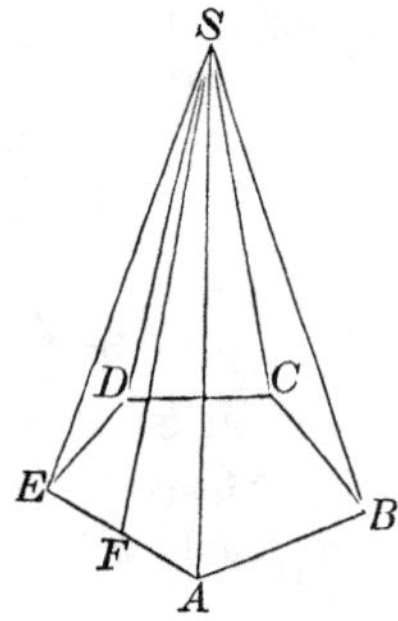

14. The *slant height* of a regular pyramid, is a line drawn from the vertex, perpendicular to one of the sides of the polygon which forms its base. Thus, *SF* is the slant height of the pyramid *S—ABCDE*.

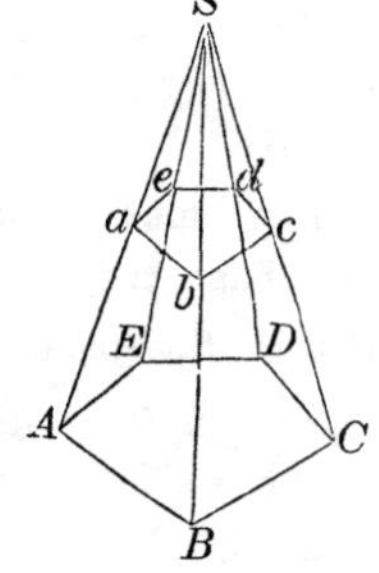

15. If from the pyramid *S—ABCDE* the pyramid S—*abcde* be cut off by a plane parallel to the base, the remaining solid, below the plane, is called the *frustum* of a pyramid.

The altitude of a frustum is the perpendicular distance between the upper and lower planes.

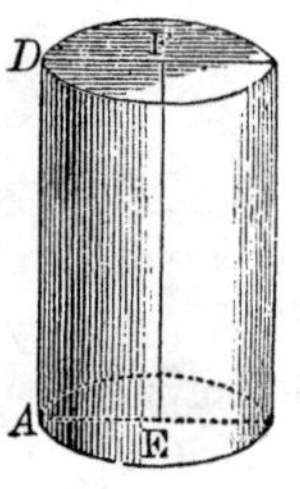

16. A *Cylinder* is a solid, described by the revolution of a rectangle, *AEFD*, about a fixed side, *EF*.

As the rectangle *AEFD*, turns around the side *EF*, like a door upon its hinges, the lines *AE* and *FD* describe circles, and the line *AD* describes the *convex surface* of the cylinder.

The circle described by the line *AE*, is called the *lower base* of the cylinder, and the circle described by *DF*, is called the *upper base*.

The immovable line EF is called the axis of the cylinder. A cylinder, therefore, is a round body with circular ends.

17. If a plane be passed through the axis of a cylinder, it will intersect it in a rectangle, PG, which is double the revolving rectangle EB.

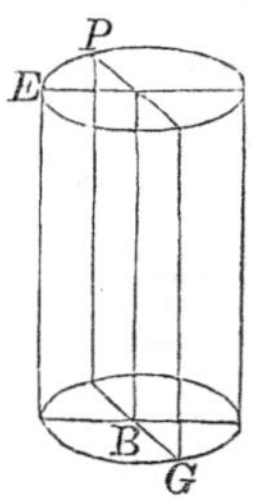

18. If a cylinder be cut by a plane parallel to the base, the section will be a circle equal to the base. For, while the side FC, of the rectangle MC, describes the lower base, the equal side MP, will describe the circle $MLKN$, equal to the lower base.

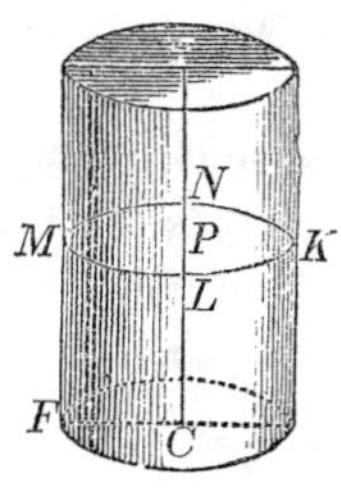

19. If a polygon be inscribed in the lower base of a cylinder, and a corresponding polygon be inscribed in the upper base, and their vertices be joined by straight lines, the prism thus formed is said to be *inscribed* in the cylinder.

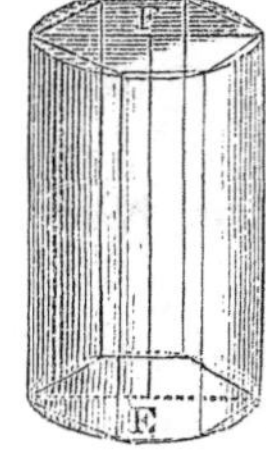

Of the Cone.

20. A *cone* is a solid, described by the revolution of a right angled triangle, ABC, about one of its sides, CB.

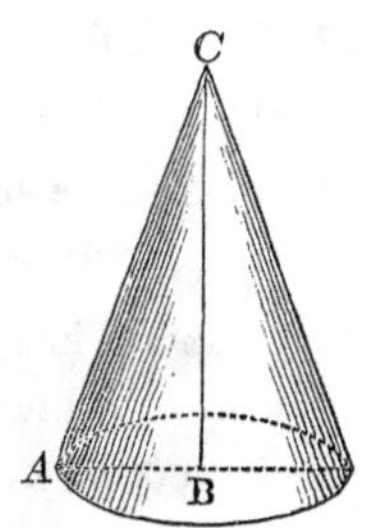

The circle described by the revolving side, AB, is called the *base* of the cone.

The hypothenuse, AC, is called the *slant height* of the cone, and the surface described by it, is called the *convex surface* of the cone.

The side of the triangle, CB, which remains fixed, is called the *axis*, or *altitude* of the cone, and the point C, the *vertex* of the cone.

21. If a cone be cut by a plane parallel to the base, the section will be a circle. For, while in the revolution of the right angled triangle SAC, the line CA describes the base of the cone, its parallel FG will describe a circle $FKHI$, parallel to the base. If from the cone $S—CDB$, the cone $S—FKH$ be taken away, the remaining part is called the *frustum* of the cone.

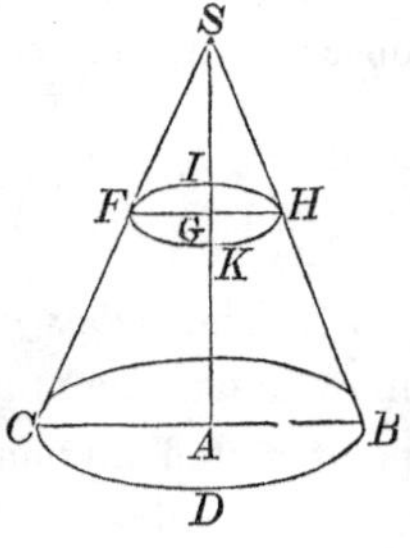

22. If a polygon be inscribed in the base of a cone, and straight lines be drawn from its vertices to the vertex of the cone, the pyramid thus formed is said to be inscribed in the cone. Thus, the pyramid $S—ABCD$ is inscribed in the cone.

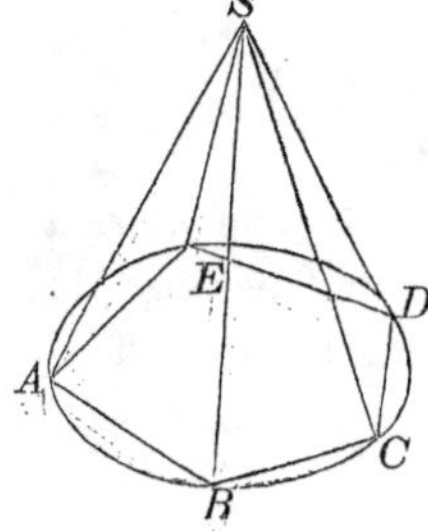

23. Two cylinders are similar, when the diameters of their bases are proportional to their altitudes.

24. Two cones are also similar, when the diameters of their bases are proportional to their altitudes.

25. A *sphere* is a solid terminated by a curved surface, all the points of which are equally distant from a certain point within called the centre.

26. The sphere may be described by revolving a semicircle, *ABD*, about the diameter *AD*. The plane will describe the solid sphere, and the semicircumference *ABD* will describe the surface

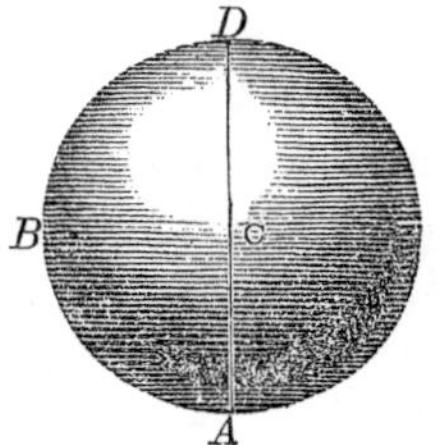

27. The *radius* of a sphere is a line drawn from the centre to any point on the circumference. Thus, *CA* is a radius.

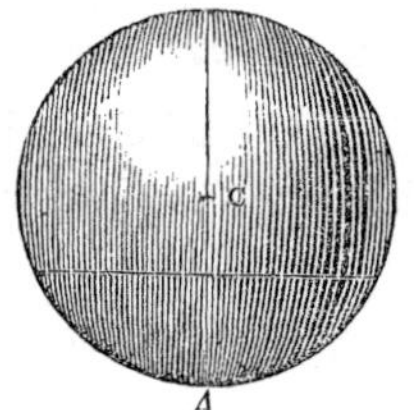

28. The *diameter* of a sphere is a line passing through the centre, and terminated by the circumference. Thus, *AD* is a diameter.

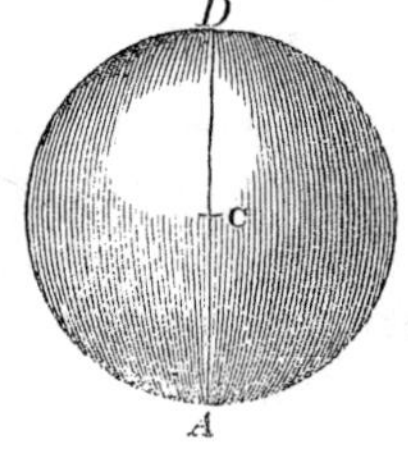

29. All diameters of a sphere are equal to each other; and each is double a radius.

30. The axis of a sphere is any line about which it revolves; and the points at which the axis meets the surface, are called the *poles*.

31. A plane is *tangent* to a sphere when it has but one point in common with it. Thus, AB is a tangent plane, touching the sphere at B.

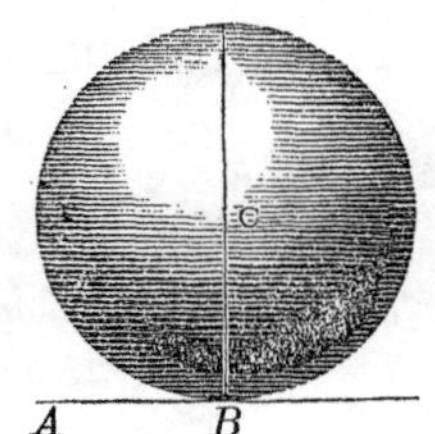

32. A *zone* is a portion of the surface of a sphere, included between two parallel planes which form its bases. Thus, the part of the surface included between the planes AE and DF is a zone. The bases of this zone are the two circles whose diameters are AE and DF.

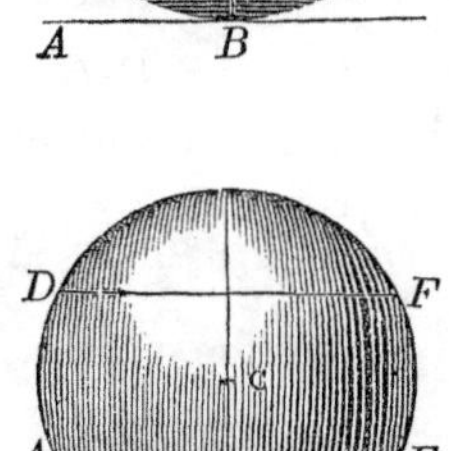

33. One of the planes which bound a zone may become tangent to the sphere; in which case the zone will have but one base. Thus, if one plane be tangent to the sphere at A, and another plane cut it in the circle DF, the zone included between them, will have but one base.

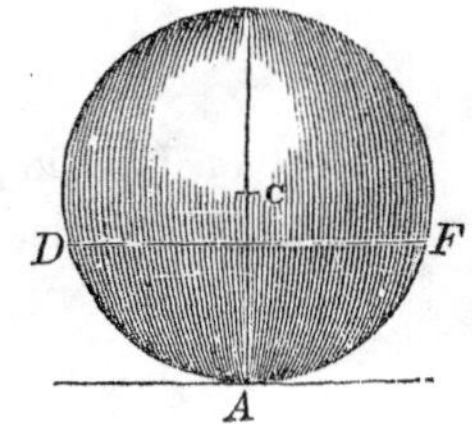

34. A *spherical segment* is a portion of the solid sphere included between two parallel planes. These parallel planes are its bases. If one of the planes is tangent to the sphere, he segment will have but one base.

35. The *altitude* of a zone or segment, is the distance between the parallel planes which form its bases.

THEOREM I.

The convex surface of a right prism is equal to the perimeter of its base multiplied by its altitude.

Let $ABCDE$—K be a right prism: then will its convex surface be equal to

$(AB+BC+CD+DE+EA)\times AF$.

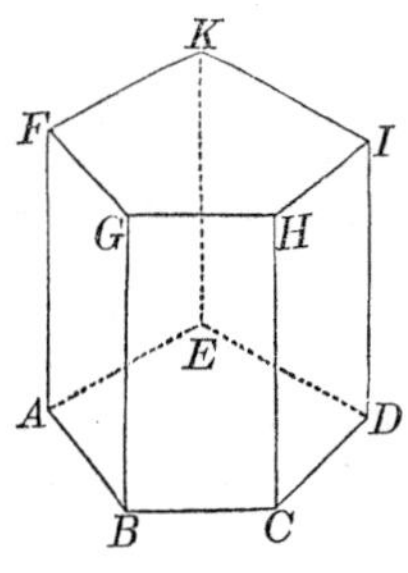

For, the convex surface is equal to the sum of the rectangles AG, BH, CI, DK, and EF, which compose it; and the area of each is equal to the product of its base by its altitude. But the altitudes are equal to the altitudes of the prism: hence, their areas, that is, the convex surface of the prism, is equal to

$$(AB+BC+CD+DE+EA)\times AF;$$

that is, equal to the perimeter of the base of the prism multiplied by its altitude.

THEOREM II.

The convex surface of a c linder is equal to the circumference of its base mu plied by its altitude.

Let DB be a cylinder, and AB the diameter of its base: the convex surface will then be equal to the altitude AD multiplied by the circumference of the base.

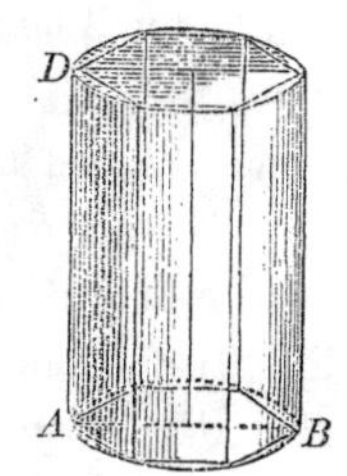

For, suppose a regular prism to be inscribed within the cylinder. Then, the convex surface of the prism will be equal to the perimeter of the base multiplied by the altitude (Th. i). But the altitude of the prism is the same as that of the cylinder; and if we suppose the sides of the polygon, which forms the base of the prism, to be indefinitely increased, the polygon will become the circle (Bk. IV. Th. xxv), in which case, its perimeter will become the circumference, and the prism will coincide with the cylinder. But its convex surface is still equal to the perimeter of its base multiplied by its altitude: hence, the convex surface of a cylinder is equal to the circumference of its base multiplied by its altitude.

THEOREM III.

In every prism the sections formed by planes parallel to the base are equal polygons.

Let AG be any prism, and IL a section made by a plane parallel to the base AC: then will the polygon IL be equal to AC.

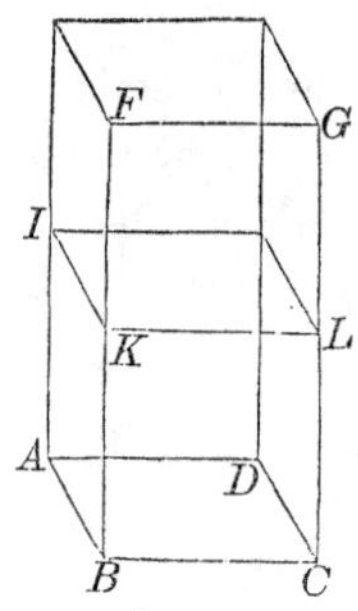

For, the two planes AC, IL, being parallel, the lines AB, IK, in which they intersect the plane AF, will also be parallel (Bk. V. Th. ix). For a like reason, BC and KL will be parallel; also, CD will be parallel to LM, and AD to IM.

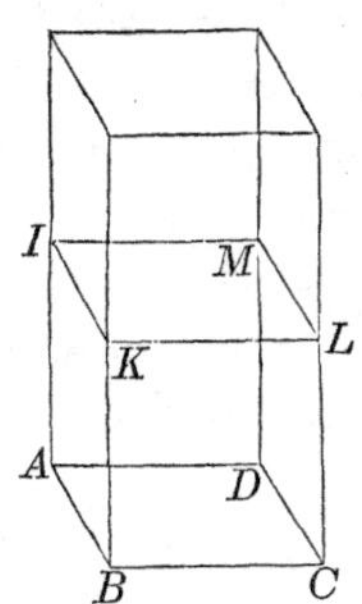

But, since AI and BK are parallel, the figure AK will be a parallelogram: hence AB is equal to IK (Bk. I. Th. xxiii). In the same way it may be shown that BC is equal to KL, CD to LM, and AD to IM.

But, since the sides of the polygon AC are respectively parallel to the sides of the polygon IL, it follows that their corresponding angles are equal (Bk. V. Th. xi), viz., the angle A to the angle I, the angle B to K, the angle C to L, and the angle M to D; hence, the polygon IL is equal to AC.

Sch. It was shown in Definition 18, that the section of a cylinder, by a plane parallel to the base, is a circle equal to the base.

THEOREM IV.

If a pyramid be cut by a plane parallel to the base,

I. *The edges and altitude will be divided proportionally.*

II. *The section will be a polygon similar to the base.*

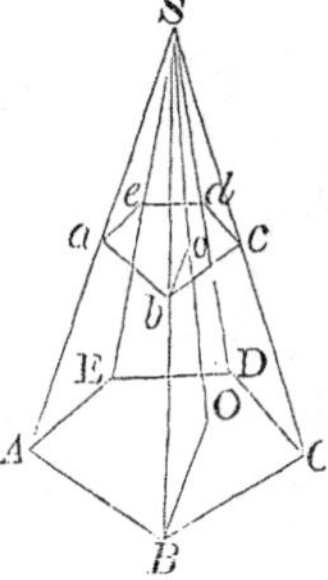

Let the pyramid $S—ABCDE$, of which SO is the altitude, be cut by the plane $abcde$ parallel to the base: then will,

$$Sa : SA :: Sb : SB,$$

and the same for the other edges; and the polygon $abcde$ will be similar to the base $ABCDE$.

First. Since the planes ABC and abc

are parallel, their intersections, AB, ab, by the plane SAB, will also be parallel (Bk. V. Th. ix); hence, the triangles SAB, sab, are similar, and we have

$$SA : Sa :: SB : Sb;$$

for a similar reason, we have

$$SB : Sb :: SC : Sc;$$

and the same for the other edges: hence, the edges SA, SB, SC, &c., are cut proportionally at the points a, b, c, &c.

The altitude SO is likewise cut proportionally at the point o; for, since BO is parallel to bo, we have

$$SO : So :: SB : Sb.$$

Secondly. Since ab is parallel to AB, bc to BC, cd to CD, &c.; the angle abc is equal to ABC, the angle bcd to BCD, and so on (Bk. V. Th. xi).

Also, by reason of the similar triangles, SAB, Sab, we have

$$AB : ab : SB : Sb,$$

and by reason of the similar triangles SBC, Sbc, we have

$$SB : Sb :: BC : bc;$$

hence (Bk. III. Th. v),

$$AB : ab :: BC : bc;$$

and for a similar reason, we also have

$$BC : bc :: CD : cd, \&c.$$

Hence, the polygons $ABCDE$, $abcde$, having their angles respectively equal, and their homologous sides proportional, are similar.

THEOREM V.

If two pyramids, having equal altitudes and their bases in the same plane, be intersected by planes parallel to the plane of the bases, the sections in each pyramid will be proportional to the bases

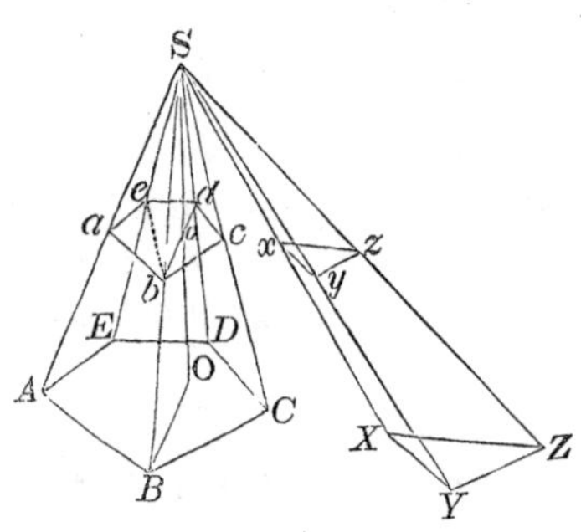

Let S—$ABCDE$, and S—XYZ, be two pyramids, having a common vertex, and their bases situated in the same plane. If these pyramids are cut by a plane parallel to the plane of their bases, giving the sections $abcde$, xyz, then will the sections $abcde$, xyz, be to each other as the bases $ABCDE$, XYZ.

For, the polygons $ABCDE$, $abcde$, being similar, their surfaces are as the squares of the homologous sides AB, ab;

but $\quad AB : ab :: SA : Sa$;

hence, $\quad ABCDE : abcde :: \overline{SA}^2 : \overline{Sa}^2$

For the same reason,

$$XYZ : xyz :: \overline{SX}^2 : \overline{Sx}^2.$$

But since abc and xyz are in one plane, the lines SA, Sa, SX, Sx, are proportional to SO, So: therefore,

$$SA : Sa :: SX : Sx:$$

hence, $\quad ABCDE : abcde :: XYZ : xyz.$

consequently, the sections $abcde$, xyz, are to each other as the bases $ABCDE$, XYZ.

Cor. If the bases $ABCDE$, XYZ, are equivalent, any sections $abcde$, xyz, made at equal distances from the bases, will be also equivalent.

THEOREM VI.

The convex surface of a regular pyramid is equal to half the product of the perimeter of its base multiplied by the slant height.

Let $S—ABCDE$ be a regular pyramid, SF its slant height: then will its convex surface be equal to half the product

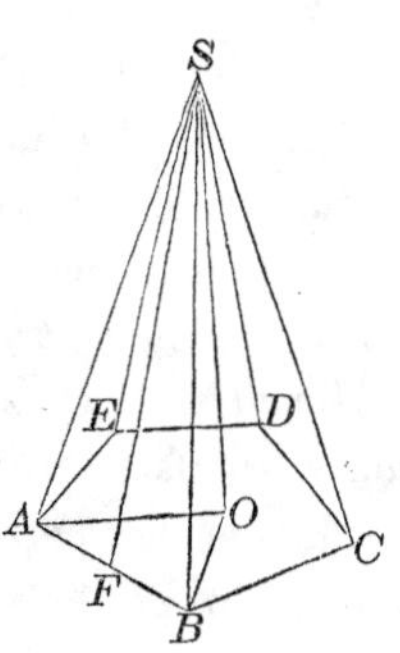

$$SF \times (AB + BC + CD + DE + EA).$$

For, since the pyramid is regular, the point O, in which the axis meets the base, is the centre of the polygon $ABCDE$; hence, the lines OA, OB, &c. drawn to the vertices of the base, are equal (Bk. IV. prob. x. Cor).

Now, in the right angled triangles SAO, SBO, the bases and perpendiculars are equal: hence, the hypothenuses are equal; and in the same way it may be proved that all the edges of the pyramid are equal. The triangles, therefore, which form the convex surface of the prism, are all equal to each other.

But the area of either of these triangles, as SAB, is equal to half the product of the base AB, by the slant height of the pyramid SF: hence, the area of all the triangles, which form the convex surface of the pyramid, is equal to half the product of the perimeter of the base by the slant height.

THEOREM VII.

The convex surface of the frustum of a regular pyramid is equal to half the sum of the perimeters of the upper and lower bases multiplied by the slant height.

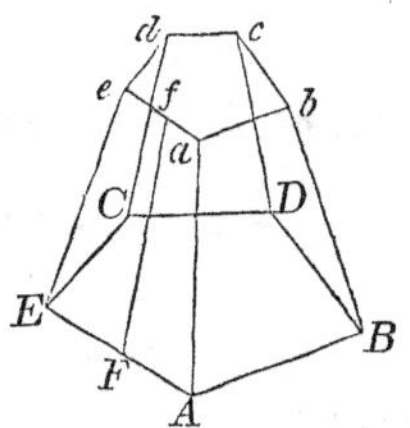

Let a—$ABCDE$ be the frustum of a regular pyramid: then will its convex surface be equal to half the product of the perimeter of its two bases multiplied by the slant height Ff.

For, since the upper base $abcde$, is similar to the lower base $ABCDE$ (Th. iv), and since $ABCDE$ is a regular polygon, it follows that the sides ab, bc, cd, de, and ea, are all equal to each other.

Hence, the trapezoids $EAae$, $ABba$, &c., which form the convex surface of the frustum are equal. But the perpendicular distance between the parallel sides of these trapezoids is equal to Ef, the slant height of the frustum.

Now, the area of either of the trapezoids, as $AEea$, is equal to half the product of $Ff \times (EA + ea)$ (Bk. IV. Th. x): hence, the area of all of them, that is, the convex surface of the frustum, is equal to half the sum of the perimeters of the upper and lower bases, multiplied by the slant height.

THEOREM VIII.

The convex surface of a cone is equal to half the product of the circumference of the base multiplied by the slant height.

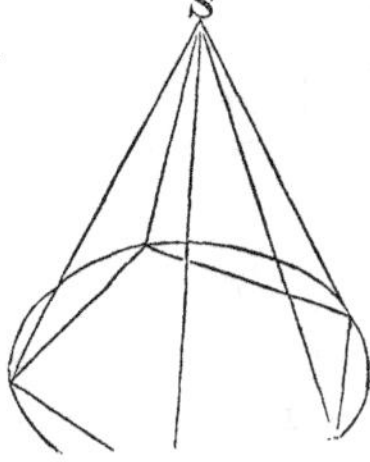

In the circle which forms the base of the cone, inscribe a regular polygon, and join the vertices with the vertex S, of the cone. We shall then have a regular pyramid inscribed in the cone.

The convex surface of this pyramid will be equal t

of the perimeter of the base by the slant height (Th. vi).

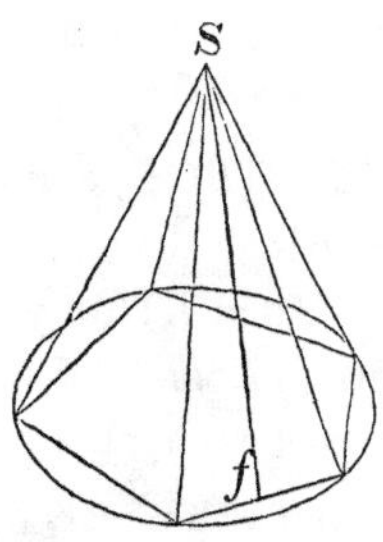

Let us now suppose the number of sides of the polygon to be indefinitely increased: the polygon will then coincide with the base of the cone, the pyramid will become the cone, and the line *Sf*, which measures the slant height of the pyramid, will then measure the slant height of the cone.

Hence, the convex surface of the cone is equal to half the product of the slant height by the circumference of the base.

THEOREM IX.

The convex surface of the frustum of a cone is equal to half the sum of the circumferences of its two bases multiplied by the slant height.

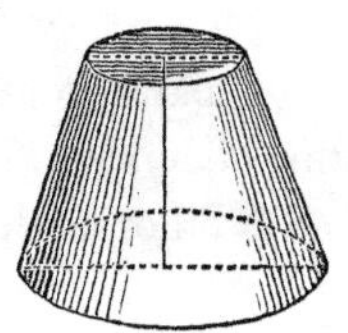

For, if we suppose the frustum of a regular pyramid to be inscribed in the frustum of a cone, its convex surface will be equal to half the product of its slant height by the perimeters of its two bases. But if we increase the number of sides of the polygons indefinitely, the frustum of the pyramid will become the frustum of the cone: hence, the area of the frustum of the cone is equal to half the sum of the circumferences of its two bases multiplied by the slant height.

THEOREM X.

Two rectangular parallelopipedons, having equal altitudes and equal bases, are equal.

Let $E—ABCD$, and $F—KGHI$, be two rectangular parallelopipedons having equal bases, AC and KH, and equal altitudes, AE and KF: then will they be equal.

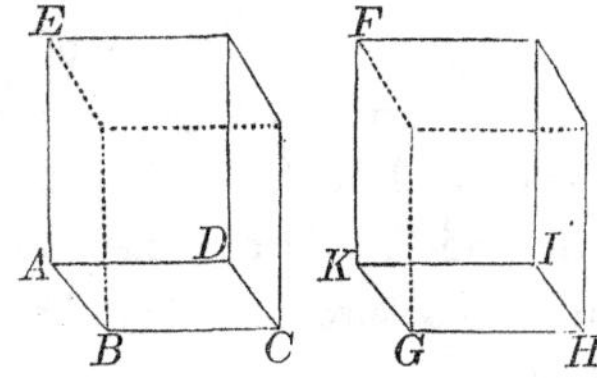

For, apply the base of the one parallelopipedon to that of the other, and since the bases are equal, they will coincide.

Again, since the edges are perpendicular to the bases, the edges of the one parallelopipedon will coincide with those of the other; and since the altitude AE is equal to KF, the planes of the upper bases will coincide. Hence, the parallelopipedons will coincide, and consequently they are equal.

THEOREM XI.

Two rectangular parallelopipedons, which have the same base, are to each other as their altitudes.

Let the parallelopipedons AG, AL, have the same base BD, then will they be to each other as their altitudes AE AI.

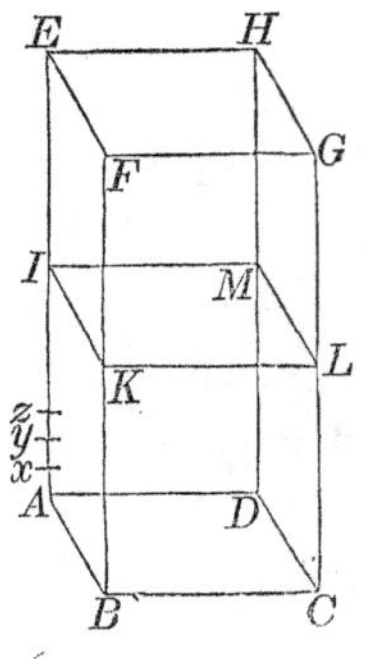

Suppose the altitudes AE, AI, to be to each other as two whole numbers, as 15 is to 8, for example. Divide AE into 15 equal parts, whereof AI will contain 8; and through x, y, z, &c., the points of division, draw planes parallel to the base. These planes will cut the solid AG into 15 partial parallelopipedons, all equal to each other, because they have equal bases and equal altitudes—equal bases, since every section, IL, made parallel to the base BD, of a prism, is equal to that base; equal altitudes, because the altitudes are the equal divisions Ax, xy, yz, &c. But of those 15 equal parallelopipedons, 8 are contained in AL;

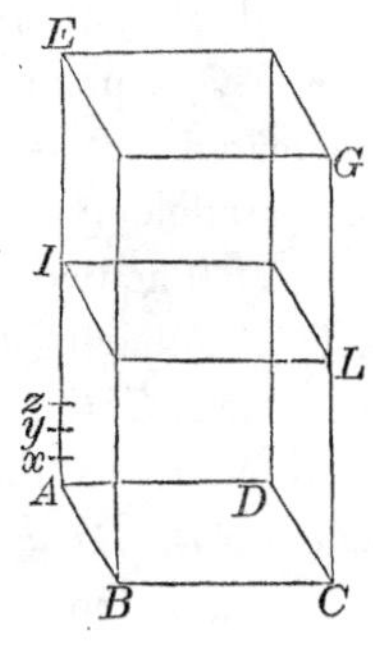

hence, *solid* AG : *solid* AL :: 15 : 8

or generally,

solid AG : *solid* AL :: AE : AI.

THEOREM XII.

Two regular parallelopipedons, having the same altitude, are to each other as their bases.

Let the parallelopipedons AG, AK, have the same altitude AE; then will they be to each other as their bases AC, AN.

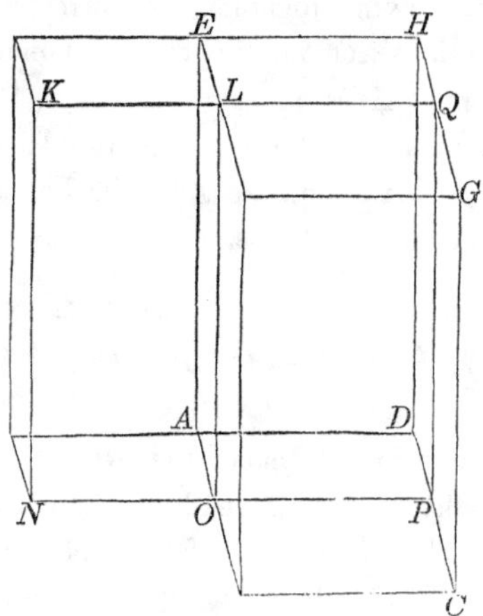

Having placed the two solids by the side of each other, as the figure represents, produce the plane $ONKL$ until it meets the plane $DCGH$ in PQ; you will thus have a third parallelopipedon AQ, which may be compared with each of the parallelopipedons AG, AK. The two solids AG, AQ, having the same base $AEHD$, are to each other as their altitudes AB, AO; in like manner, the two solids AQ AK, having the same base $AOLE$, are to each other as their altitudes AD, AM.

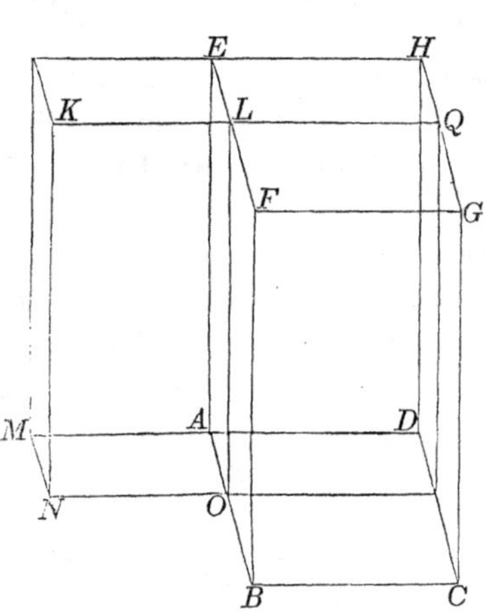

Hence, we have the two proportions,

$$\text{solid } AG : \text{solid } AQ :: AB : AO,$$
$$\text{solid } AQ : \text{solid } AK :: AD : AM.$$

Multiplying together the corresponding terms of these proportions, and omitting the common multiplier *solid* AQ, we have

$$\text{solid } AG : \text{solid } AK :: AB \times AD : AO \times AM.$$

But $AB \times AD$ represents the base $ABCD$; and $AO \times AM$ represents the base $AMNO$: hence, two rectangular parallelopipedons of the same altitude are to each other as their bases

THEOREM XIII.

Any two rectangular parallelopidedons are to each other as the products of their three dimensions.

For, having placed the two solids AG, AZ, (see next figure) so that their surfaces have the common angle BAE, produce the planes necessary for completing the third parallelopipedon AK, having the same altitude with the parallelopipedon AG By the last proposition we shall have the proportion, solid AG : solid AK :: $ABCD$: $AMNO$.

But the two parallelopipedons AK, AZ, having the same base $AMNO$, are to each other as their altitudes AE, AX; hence, we have

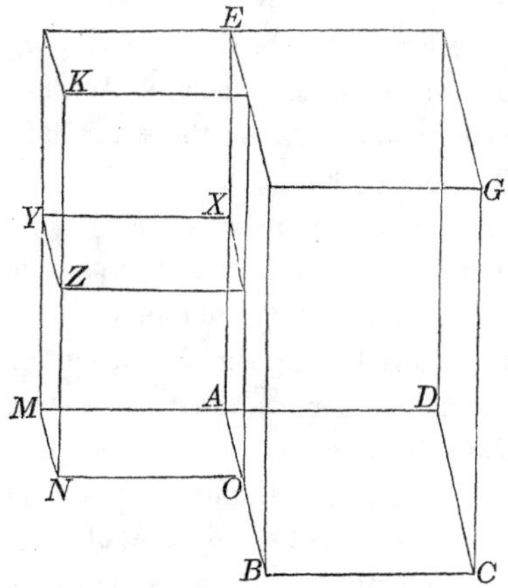

solid AK : solid AZ :: AE : AX.

Multiplying together the corresponding terms of these proportions, and omitting in the result the common multiplier solid AK, we shall have

solid AG : solid AZ :: $ABCD \times AE$: $AMNO \times AX$.

Instead of the bases $ABCD$ and $AMNO$, put $AB \times AD$ and $AO \times AM$, and we have

solid AG : solid AZ :: $AB \times AD \times AE$: $AO \times AM \times AX$.

Hence, any two rectangular parallelopipedons are to each other as the product of their three dimensions.

Sch. We are consequently authorized to assume, as the measure of a rectangular parallelopipedon, the product of its three dimensions.

In order to comprehend the nature of this measurement, it is necessary to reflect, that the number of linear units in one dimension of the base multiplied by the number of linear units of the other dimension of the base, will give the number of superficial units in the base of the parallelopipedon (Bk. IV. Th. vi. Sch). For each unit in height, there are evidently as many solid units as there are superficial units in the base Therefore, the number of superficial units in the base multiplied by the number of linear units in the altitude, gives the number of solid units in the parallelopipedon.

If the three dimensions of another parallelopipedon are valued according to the same linear unit, and multiplied together in the same manner, the two products will be to each other as the solids, and will serve to express their relative magnitude.

Let us illustrate this by an example.

Let $ABCD$ be the base of a parallelopipedon, and suppose $AB=4$ feet, and $BC=3$ feet. Then the number of square feet in the base $ABCD$ will be equal to $3\times4=12$ square feet.

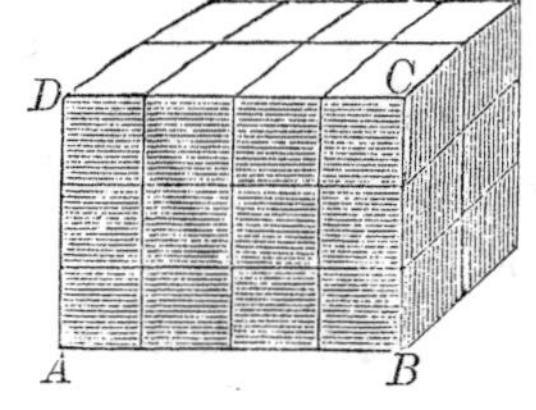

Therefore, 12 equal cubes of 1 foot each, may be placed by the side of each other on the base. If the parallelopipedon be 1 foot in height, it will contain 12 cubic feet; were it 2 feet in height, it would contain two tiers of cubes, or 24 cubic feet; were it 3 feet in height, it would contain three tiers of cubes, or 36 cubic feet.

The magnitude of a solid, its volume or extent, forms what is called its *solidity;* and this word is exclusively employed to designate the measure of a solid; thus, we say the solidity of a rectangular parallelopipedon is equal to the product of its base by its altitude, or to the product of its three dimensions.

As the cube has all its three dimensions equal, if the side is 1, the solidity will be $1 \times 1 \times 1 = 1$; if the side is 2, the solidity will be $2 \times 2 \times 2 = 8$; if the side is 3, the solidity will be $3 \times 3 \times 3 = 27$; and so on: hence, if the sides of a series of cubes are to each other as the numbers 1, 2, 3, &c. the cubes themselves, or their solidities, will be as the numbers 1, 8, 27, &c. Hence it is, that in arithmetic, the *cube* of a number is the name given to a product which results from three factors, each equal to this number.

THEOREM XIV.

If a parallelopipedon, a prism, and a cylinder, have equivalent bases and equal altitudes, they will be equivalent.

Let F—$ABCD$, be a parallelopipedon; F—$ABCDE$, a prism; and D—ABC, a cylinder, having equivalent bases and equal altitudes: then will they be equivalent.

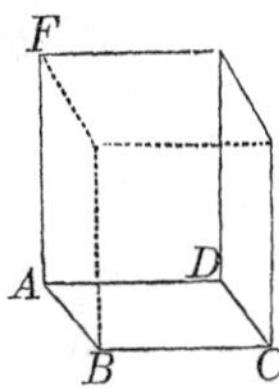

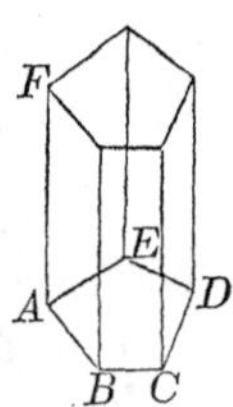

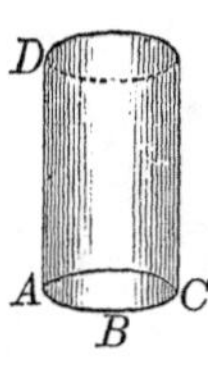

For, since their bases are equivalent they will contain the same number of units of surface (Bk. IV. Def. 9). Now, for each unit of height there will be one tier of equal cubes in each solid, and since the altitudes are equal, the number of tiers in each solid will be equal: hence, the solidities will be equal, and therefore the solids will be equivalent.

Cor. Hence, we conclude, that the solidity of a prism or cylinder is equal to the area of its base multiplied by its altitude.

THEOREM XV.

Two triangular pyramids, having equivalent bases and equal altitudes, are equivalent, or equal in solidity.

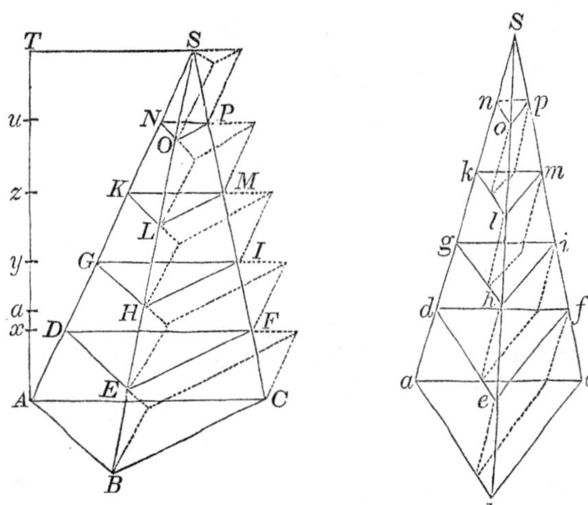

Let their equivalent bases, *ABC*, *abc*, be situated in the same plane, and let *AT* be their common altitude. If they are not equivalent, let *S—abc* be the smaller; and suppose *Aa* to be the altitude of a prism, which, having *ABC* for its base, is equal to their difference.

Divide the altitude *AT* into equal parts *Ax*, *xy*, *yz*, &c., each less than *Aa*, and let *k* be one of those parts: through the points of division pass planes parallel to the plane of the bases: the corresponding sections formed by these planes in the two pyramids will be respectively equivalent, namely, *DEF* to *def*, *GHI* to *ghi*, &c. (Th. v. Cor.).

This being granted, upon the triangles *ABC*, *DEF*, *GHI*, &c., taken as bases, construct exterior prisms having for edges the parts *AD*, *DG*, *GK*, &c., of the edge *SA*; in like manner, on bases *def*, *ghi*, *klm*, &c., in the second pyramid construct interior prisms, having for edges the corresponding parts of *Sa*. It is plain that the sum of the exterior prisms of the pyramid *S—ABC* will be greater than the pyramid; while the sum of the interior prisms of the pyramid *S—abc*, will be less than the pyramid. Hence, the difference between these sums will be greater than the difference between the pyramids.

Now, beginning with the bases *ABC*, *abc*, the second exterior prism *DEF—G* is equivalent to the first interior prism *def—a*, because they have the same altitude *k*, and their bases *DEF*, *def*, are equivalent; for like reasons, the third exterior prism *GHI—K*, and the second interior prism *ghi—d*, are equivalent; the fourth exterior and the third interior; and so on, to the last of each series. Hence, all the exterior prisms of the pyramid *S—ABC*, excepting the first prism *ABC—D*, have equivalent corresponding ones in the interior prisms of the pyramid *S—abc*: hence, the prism *ABC—D* is the difference between the sum of all the exterior prisms of the pyramid *S—ABC*, and of the interior prisms of the pyramid *S—abc*. But this difference has already been proved to be greater than that of the two pyramids: which, by supposition, differ by the prism *a—ABC*: hence, the prism *ABC—D*, must be greater than the prism *a—ABC*. But in reality it is less, for they have the same base *ABC*, and the altitude *Ax*, of the first, is less than *Aa*, the altitude of the second. Hence, the supposed inequality between the two pyramids cannot exist: hence, the two pyramids; *S—ABC*, *S—abc*, having equal altitudes and equivalent bases, are themselves equivalent.

THEOREM XVI.

Every triangular pyramid is a third part of a triangular prism having the same base and the same altitude.

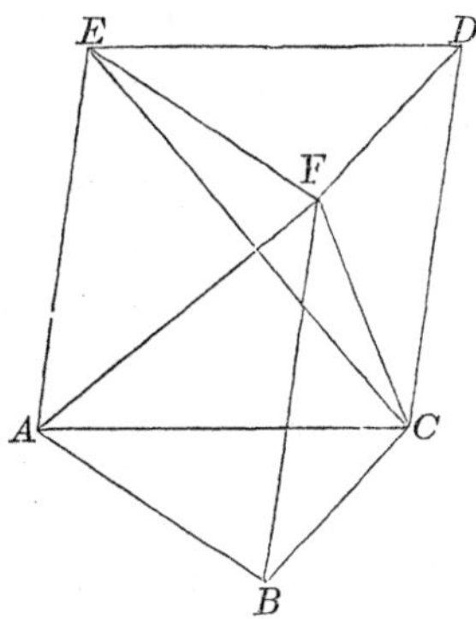

Let *F—ABC* be a triangular pyramid, *ABC—DEF* a triangular prism of the same base and the same altitude: the pyramid will be equal to a third of the prism.

Cut off the pyramid *F—ABC* from the prism, by the plane *FAC;* there will remain the solid *F—ACDE,* which may be considered as a quadrangular pyramid, whose vertex is *F*, and whose base is the parallelogram *ACDE.* Draw the diagonal *CE;* and pass the plane *FCE,* which will cut the quadrangular pyramid into two triangular ones, *F-ACE, F-CDE.* These two triangular pyramids have for their common altitude the perpendicular let fall from *F* on the plane *ACDE;* and their bases are also equal, being halves of the parallelogram *AD:* hence, the pyramid *F-ACE,* and the pyramid *F-CDE,* are equivalent (Th. xv).

But the pyramid *F—CDE,* and the pyramid *F—ABC,* have equal bases, *ABC, DEF;* they have also the same altitude, namely, the distance between the parallel planes *ABC, DEF,* hence, the two pyramids are equivalent. Now, the pyramid *F—CDE* has already been proved equivalent to *F—ACE;* hence, the three pyramids *F—ABC, F—CDE, F—ACE,* which compose the prism *ABC—DEF* are all equivalent.

Hence, the pyramid $F—ABC$ is the third part of the prism $ABC—DEF$, which has the same base and the same altitude.

Cor. The solidity of a triangular pyramid is equal to a third part of the product of its base by its altitude.

THEOREM XVII.

The solidity of every pyramid is equal to the base multiplied by a third of the altitude.

Let $S—ABCDE$ be a pyramid.

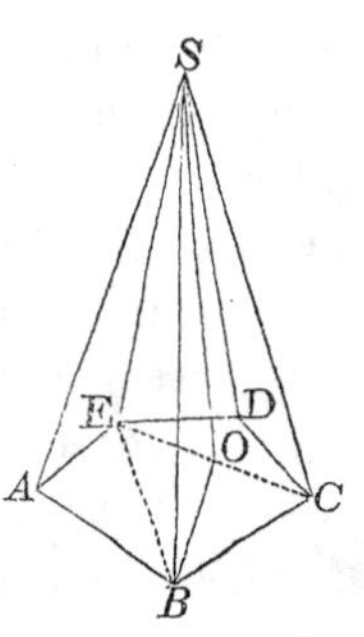

Pass the planes SEB, SEC through the diagonals EB, EC; the polygonal pyramid $S—ABCDE$ will be divided into several triangular pyramids all having the same altitude SO. But each of these pyramids is measured by multiplying its base ABE, BCE, or CDE, by the third part of its altitude SO (Th. xvi. Cor.); hence the sum of these triangular pyramids, or the polygonal pyramid $S—ABCDE$, will be measured by the sum of the triangles ABE, BCE, CDE, or the polygon $ABCDE$, multiplied by one third of SO.

Cor. 1. Every pyramid is the third part of the prism which has the same base and the same altitude.

Cor. 2. Two pyramids having the same altitude, are to each other as their bases.

Cor. 3. Two pyramids having equivalent bases, are to each other as their altitudes.

Cor. 4. Pyramids are to each other as the products of their bases by their altitudes

THEOREM XVIII.

The solidity of a cone is equal to one third of the product of the base multiplied by the altitude.

Let $ABCDE$ be the base, S the vertex, and SO the altitude of the cone: then will its solidity be equal to one third the product of its base by its altitude SO.

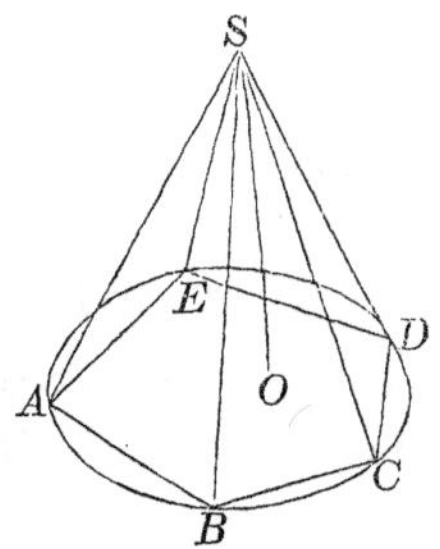

Inscribe in the base of the cone any regular polygon, $ABCDE$, and join the vertices A, B, C, &c., with the vertex S, of the cone; then will there be inscribed in the cone a regular pyramid, having for its base the polygon $ABCDE$. The solidity of this pyramid is equal to one third of the base multiplied by the altitude (Th. xvii).

Let now, the number of sides of the polygon be indefinitely increased: the polygon will then become equal to the circle, and the pyramid and cone will coincide and become equal. But the solidity of the pyramid will still be equal to one third of the product of the base multiplied by the altitude, whatever be the number of sides of the polygon which forms its base: hence, the solidity of the cone is equal to one third of the product of its base multiplied by its altitude.

Cor. 1. A cone is the third part of a cylinder having the same base and the same altitude; whence it follows:

1st, That cones of equal altitudes are to each other as their bases.

2nd, That cones of equal bases are to each other as their altitudes

Cor. 2. The solidity of a cone is equivalent to the solidity of a pyramid having an equivalent base and the same altitude.

THEOREM XIX.

Similar prisms are to each other as the cubes of their homologous edges.

Let ABC--D, EFG--H be similar prisms: then we shall have

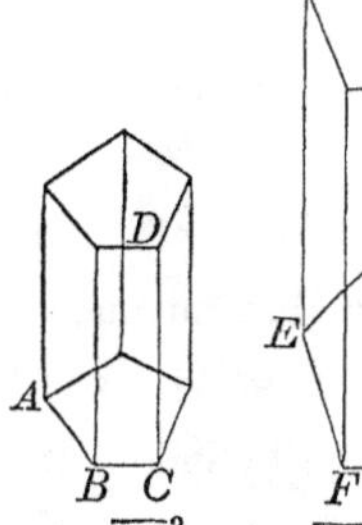

$$solid\ AD : solid\ EH :: \overline{AB}^3 : \overline{EF}^3;$$

or $$solid\ AD : solid\ EH :: \overline{CD}^3 : \overline{HG}^3;$$

or, the solids will be to each other as the cubes of any other of their homologous edges.

For, the solids are to each other as the products of their bases and altitudes (Th. xiv. Cor.), that is,

$$solid\ ABC\text{–}D : solid\ EFG\text{–}H :: ABC \times CD : EFG \times GH.$$

But the bases being similar polygons are to each other as the squares of their like sides (Bk. IV. Th. xxi); that is,

$$ABC : EFG :: \overline{AB}^2 : \overline{EF}^2,$$

therefore,

$$solid\ ABC\text{–}D : solid\ EFG\text{–}H :: \overline{AB}^2 \times CD : \overline{EF}^2 \times GH.$$

But since the solids are similar, the parallelograms BD and FH are similar (Def. 3): hence, CD and GH are proportional to BC and FG, and consequently to AB and EF: hence, we have,

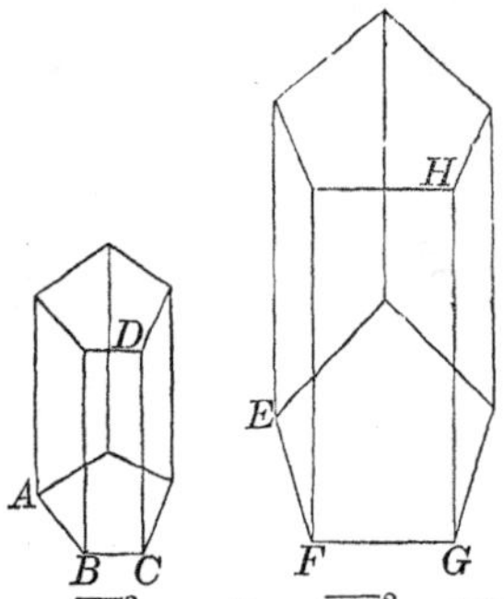

$$\text{solid } ABC\text{–}D : \text{solid } EFG\text{–}H :: \overline{AB}^2 \times AB : \overline{EF}^2 \times EF.$$

that is,

$$\text{solid } ABC\text{–}D : \text{solid } EFG\text{–}H :: \overline{AB}^3 : \overline{EF}^3;$$

and in a similar manner it may be shown that the solids are to each other as the cubes of any other homologous sides.

Cor. Since cylinders are to each other as the product of their bases and altitudes (Th. xiv. Cor.), it follows that similar cylinders are to each other as the cubes of the linear dimensions.

THEOREM XX.

Every section of a sphere, made by a plane, is a circle.

Let AMB be a section, made by a plane, in the sphere whose centre is C.

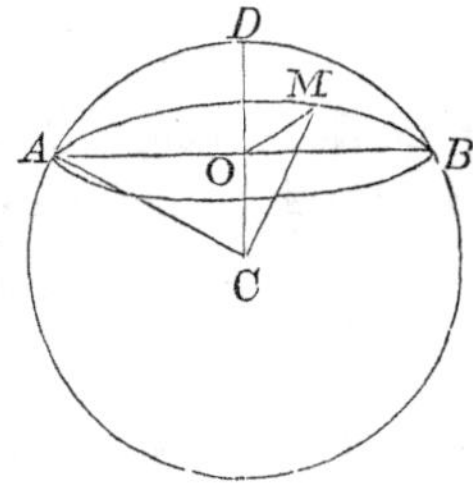

From the centre C draw CO, perpendicular to the plane AMB, and also draw the lines CA, CM, &c., to the points of the curve AMB, which terminate the section, and join OA, OM, &c.

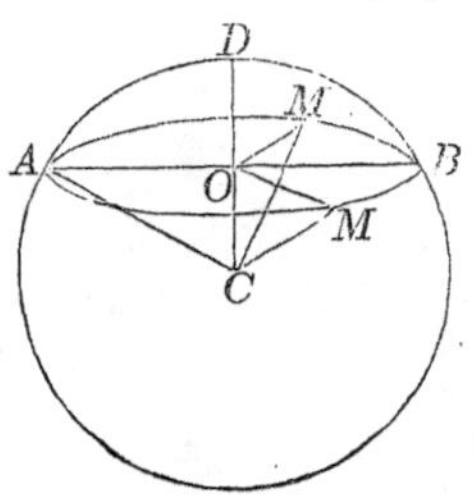

Then, since CO is perdendicular to the plane AMB, the angles COA, COM &c., will be right angles, and since the radii of the sphere are all equal, the right angled triangles CAO, COM, &c., will have the hypothenuses equal, and the side CO common: hence, the remaining sides will be equal (Bk. I. Th. xix). Therefore, all lines drawn from O to any point of the curve AMB are equal: hence AMB is a circle.

Cor. 1. If the section passes through the centre of the sphere, its radius will be the radius of the sphere: hence, all great circles are equal.

Cor. 2. Two great circles always bisect each other; for their common intersection, passing through the centre, is a diameter.

Cor. 3. Every great circle divides the sphere and its surface into two equal parts: for, if the two hemispheres were separated and afterwards placed on the common base, with their convexities turned the same way, the two surfaces would exactly coincide, no point of the one being nearer the centre than any point of the other.

Cor. 4. The centre of a small circle, and that of the sphere, are in the same straight line, perpendicular to the plane of the small circle.

Cor. 5. Small circles are the less the farther they lie from the centre of the sphere ; for the greater CO is, the less is the chord AB, the diameter of the small circle AMB.

THEOREM XXI.

Every plane perpendicular to a radius at its extremity is tangent to the sphere.

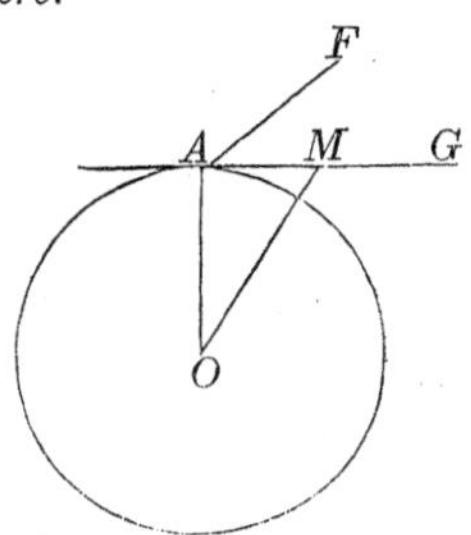

Let FAG be a plane perpendicular to the radius OA, at its extremity A. Any point M, in this plane, being assumed, and OM, AM, being drawn, the angle OAM will be a right angle, and hence, the distance OM will be greater than OA. Hence, the point M lies without the sphere ; and as the same can be shown for every other point of the plane FAG, this plane can have no point but A common to it and the surface of the sphere ; hence it is a tangent plane (Def. 31).

Sch. In the same way it may be shown, that two spheres have but one point in common, and therefore touch each other, when the distance between their centres is equal to the sum, or the difference of their radii ; in either case, the centres and the point of contact lie in the same straight line.

THEOREM XXII.

If a regular semi-polygon be revolved about a line passing through the centre and the vertices of two opposite angles, the surface described by its perimeter will be equal to the axis multiplied by the circumference f the inscribed circle.

Suppose the regular semi-polygon *ABCDE* to be revolved about the line *AF* as an axis: then will the surface described by its perimeter be equal to *AF* multiplied by the circumference of the inscribed circle.

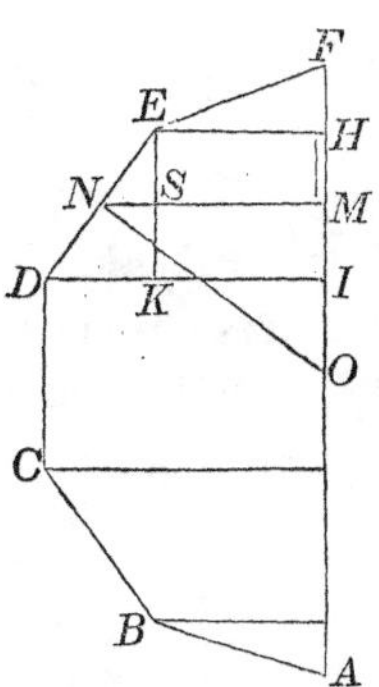

From *E* and *D*, the extremities of one of the equal sides, let fall the perpendiculars *EH*, *DI*, on the axis *AF*, and from the centre *O*, draw *ON* perpendicular to the side *DE*: *ON* will then be the radius of the inscribed circle (Bk. IV. Prob. x).

Let us first find the measure of the surface described by one of the equal sides, as *DE*.

From *N*, the middle point of *DE*, draw *NM* perpendicular to the axis *AF*, and through *E*, draw *EK*, parallel to it, meeting *MN* in *S*.

Then, since *EN* is half of *ED*, *NS* will be half of *DK* (Bk. IV. Th. xiii): and hence, *NM* is equal to half the sum of $EH+DI$.

But, since the circumferences of circles are to each other as their diameters (Bk. IV. Th. xxiv), or as their radii, the halves of the diameters, we shall have the circumference described by the point *N*, equal to half the sum of the circumferences described by the points *D* and *E*.

But in the revolution of the polygon the line *ED* describes the surface of the frustum of a cone, the measure of which is equal to *DE* multiplied into half the sum of the circumferences of the two bases (Th. ix); that is, equal to *DE* into the circumference described by the point *N*.

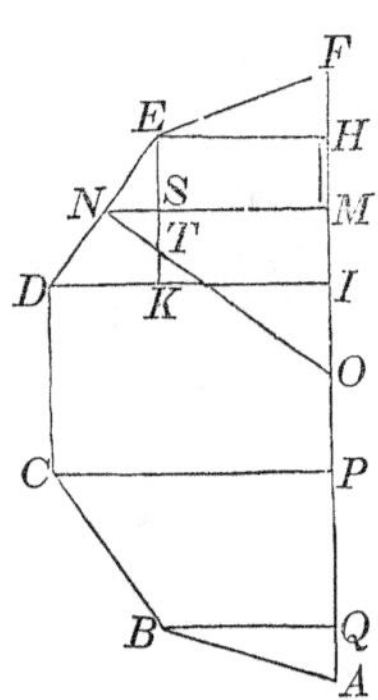

But, the triangle ENS is similar to SNT (Bk. IV. Th. xviii), and also to EDK, and since TNS is similar to ONM, it follows that EDK and ONM are similar; hence,

$$ED : EK \text{ or } HI :: ON : NM,$$

or $ED : HI :: \textit{circumference } ON : \textit{circumference } MN.$

consequently,

$$ED \times \textit{circumference } MN = HI \times \textit{circumference } ON,$$

that is, ED multiplied into the circumference of the circle described with the radius NM, is equal to HI into the circumference of the circle described with the radius ON. But the former is equal to the surface described by the line ED in the revolution of the polygon about the axis AF; hence, the latter is equal to the same area; and since the same may be shown for each of the other sides, it is plain that the surface described by the entire perimeter is equal to

$$(FH+HI+IP+PQ+QA) \times \textit{cir'f. } ON = AF \times \textit{cir'f. } ON.$$

Cor. The surface described by any portion of the perimeter, as EDC, is equal to the distance between the two perpendiculars let fall from its extremities, on the axis, multiplied by the circumference of the inscribed circle. For, the surface described by DE is equal to $HI \times$ circumference ON, and the surface described by DC is equal to $IP \times$ circumfe rence ON: hence, the surface described by $ED+DC$, is equal to $(HI+IP)\times$ circumference ON, or equal to $HP\times$ circumference ON.

THEOREM XXIII.

The surface of a sphere is equal to the product of its diameter by the circumference of a great circle.

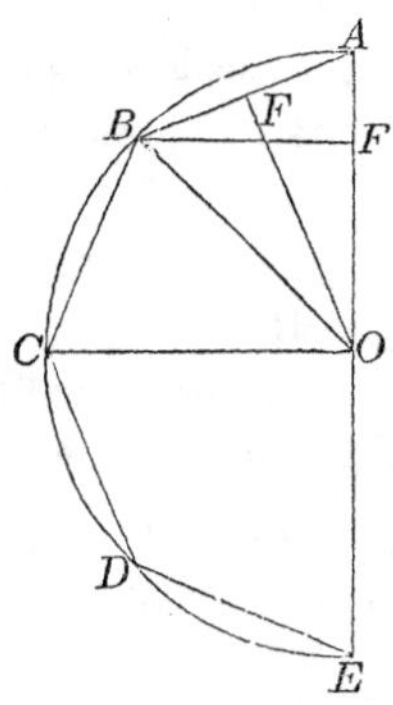

Let $ABCDE$ be a semicircle. Inscribe in it any regular semi-polygon, and from the centre O draw OF perpendicular to one of the sides.

Let the semicircle and the semi-polygon be revolved about the axis AE: the semicircumference $ABCDE$ will describe the surface of a sphere (Def. 26); and the perimeter of the semi-polygon will describe a surface which has for its measure $AE\times$ circumference OF (Th. xxii); and this will be true whatever be the number of sides of the polygon. But if the number of sides of the polygon be indefinitely increased, its perimeter will coincide with the circumference $ABCDE$, the perpendicular OF will become equal to OE, and the surface described by the perimeter of the semi-polygon will then be the same as that described by the semicircumference $ABCDE$. Hence, the surface of the sphere is equal to $AE\times$ circumference OE.

Cor. Since the area of a great circle is equal to the product of its circumference by half the radius, or by one-fourth of the diameter (Bk. IV. Th. xxvii), it follows that the surface of a sphere is equal to four of its great circles.

THEOREM XXIV.

The surface of a zone is equal to its altitude multiplied by the circumference of a great circle.

For, the surface described by any portion of the perimeter of the inscribed polygon, as $BC+CD$ is equal to $EH\times$ circumference OF (Th. xxii. Cor). But when the number of sides of the polygon is indefinitely increased, $BC+CD$, becomes the arc BCD, OF becomes equal to OA, and the surface described by $BC+CD$, becomes the surface of the zone described by the arc BCD: hence, the surface of the zone is equal to $EH\times$ circumference OA.

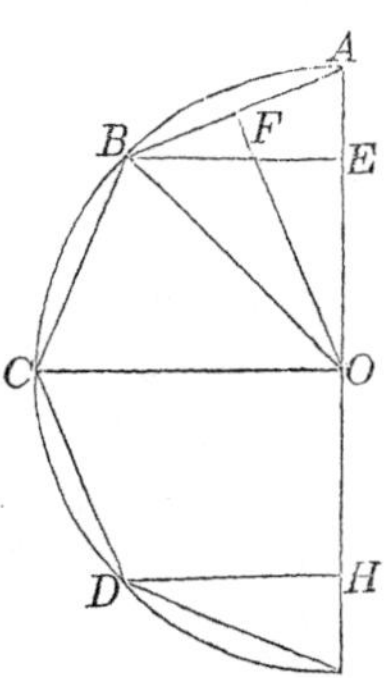

Sch. 1. When the zone has but one base, as the zone described by the arc $ABCD$, its surface will still be equal to the altitude AE multiplied by the circumference of a great circle.

Sch. 2. Two zones taken in the same sphere, or in equal spheres, are to each other as their altitudes; and any zone is to the surface of the sphere as the altitude of the zone is to the diameter of the sphere.

THEOREM XXV.

The solidity of a sphere is equal to one third of the product of the surface multiplied by the radius.

For, conceive a polyedron to be inscribed in the sphere.

This polyedron may be considered as formed of pyramids, each having for its vertex the centre of the sphere, and for its base one of the faces of the polyedron. Now, the solidity of each pyramid, will be equal to one third of the product of its base by its altitude (Th. xvii).

But if we suppose the faces of the polyedron to be continually diminished, and consequently, the number of the pyramids to be constantly increased, the polyedron will finally become the sphere, and the bases of all the pyramids will become the surface of the sphere. When this takes place, the solidities of the pyramids will still be equal to one third the product of the bases by the common altitude, which will then be equal to the radius of the sphere.

Hence, the solidity of a sphere is equal to one third of the product of the surface by the radius.

THEOREM XXVI.

The surface of a sphere is equal to the convex surface of the circumscribing cylinder; and the solidity of the sphere is two thirds the solidity of the circumscribing cylinder.

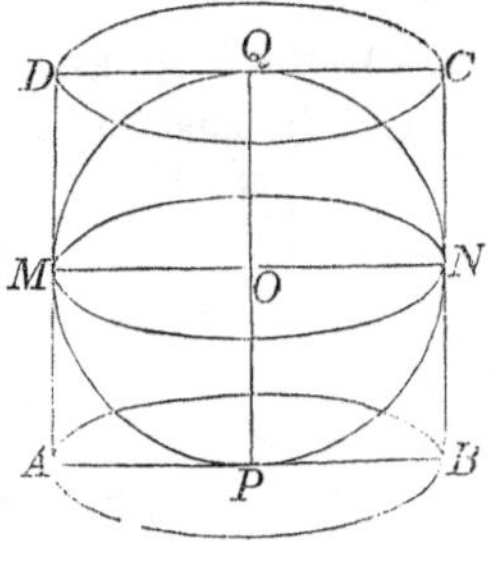

Let $MPNQ$ be a great circle of the sphere; $ABCD$ the circumscribing square: if the semicircle PMQ, and the half square $PADQ$, are at the same time made to revolve about the diameter PQ, the semicircle will describe the sphere, while the half square will describe the cylinder circumscribed about that sphere.

The altitude AD, of the cylinder, is equal to the diameter

PQ; the base of the cylinder is equal to the great circle, since its diameter AB is equal MN; hence, the convex surface of the cylinder is equal to the circumference of the great circle multiplied by its diameter (Th. ii). This measure is the same as that of the surface of the sphere (Th. xxiii): hence, the surface of the sphere is equal to the convex surface of the circumscribing cylinder.

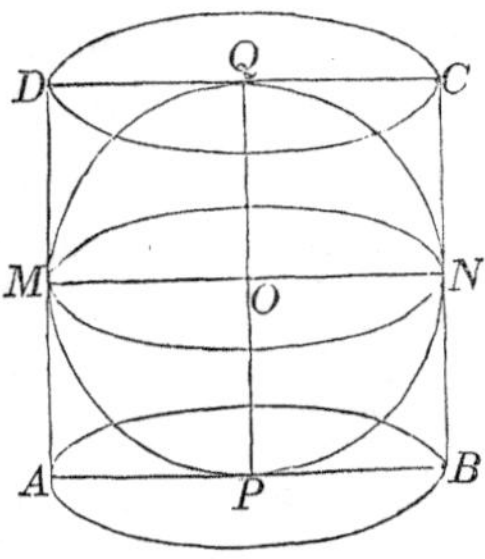

In the next place, since the base of the circumscribing cylinder is equal to a great circle, and its altitude to the diameter, the solidity of the cylinder will be equal to a great circle multiplied by a diameter (Th. xiv. Cor). But the solidity of the sphere is equal to its surface multiplied by a third of its radius; and since the surface is equal to four great circles (Th. xxiii. Cor.), the solidity is equal to four great circles multiplied by a third of the radius; in other words, to one great circle multiplied by four-thirds of the radius, or by two-thirds of the diameter; hence, the sphere is two-thirds of the circumscribing cylinder.

APPENDIX

OF THE FIVE REGULAR POLYEDRONS.

A *regular polyedron,* is one whose faces are all equal polygons, and whose solid angles are equal. There are five such solids.

1. The *Tetraedron*, or equilateral pyramid, is a solid bounded by four equal triangles.

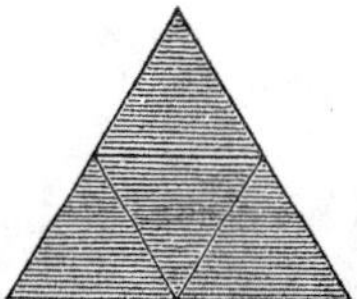

2. The *hexaedron* or *cube*, is a solid, bounded by six equal squares.

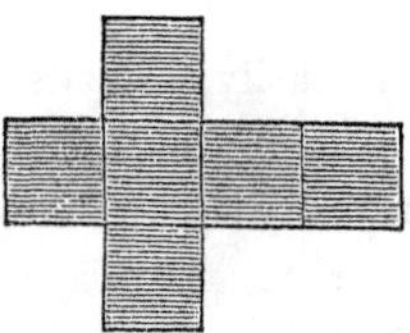

3. The *octaedron*, is a solid, bounded by eight equal equilateral triangles.

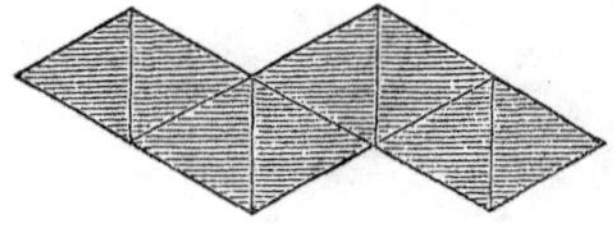

4. The *dodecaedron*, is a solid bounded by twelve equal pentagons.

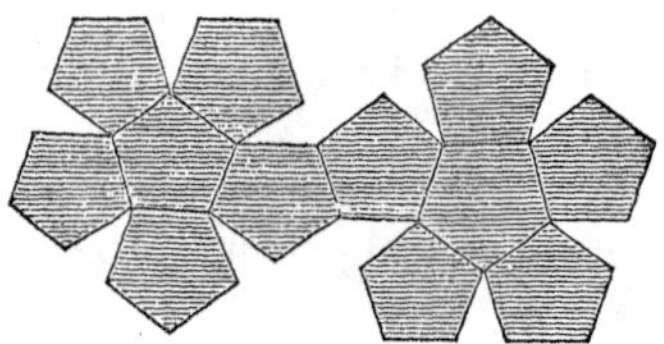

5. The *icosaedron*, is a solid, bounded by twenty equal equilateral triangles.

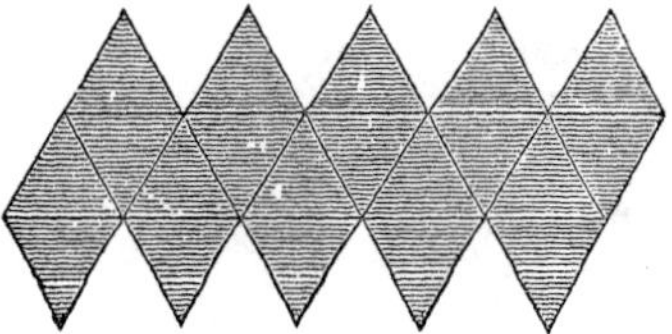

6. The regular solids may easily be made of pasteboard.

Draw the figures of the regular solids accurately on pasteboard, and then cut through the bounding lines: this will give figures of pasteboard similar to the diagrams. Then, cut the other lines half through the pasteboard, after which, turn up the parts, and glue them together, and you will form the bodies which have been described.

APPLICATIONS

OF

GEOMETRY.

MENSURATION OF SURFACES.

DEFINITIONS.

1. The area of any figure has already been defined to be the measure of its surface (Bk. IV. Def. 7). This measure is merely the number of squares which the figure is equal to.

A square whose side is one inch, one foot, or one yard, &c., is called the *measuring unit;* and the area or contents of a figure is expressed by the number of such squares which the figure contains.

2. In the questions involving decimals, the decimals are generally carried to four places, and then taken to the nearest figure. That is, if the fifth decimal figure is 5, or greater than 5, the fourth figure is increased by one.

3. Surveyors, in measuring land, generally use a chain called Gunter's chain. This chain is four rods, or 66 feet in length, and is divided into 100 links.

4. An *acre* is a surface equal in extent to 10 square chains; that is, equal to a rectangle of which one side is ten chains, and the other side one chain.

One quarter of an acre, is called a *rood.*

Since the chain is 4 rods in length, 1 square chain contains 16 square rods ; and therefore, an acre, which is 10 square chains, contains 160 square rods, and a rood contains 40 square rods. The square rods are called perches.

5. Land is generally computed in acres, roods, and perches, which are respectively designated by the letters A, R, P.

When the linear dimensions of a survey are chains or links, the area will be expressed in square chains or square links, and it is necessary to form a rule for reducing this area to acres, roods, and perches. For this purpose, let us form the following

TABLE.

1 square chain $=100\times100=10000$ square links.
1 acre $=10$ square chains $=100000$ square links.
1 acre $=4$ roods $=160$ perches.
1 square mile $=6400$ square chains $=640$ acres.

6. Now, when the linear dimensions are lines, the area will be expressed in square links, and may be reduced to acres by dividing by 100000, the number of square lines in an acre: that is, by pointing off five decimal places from the right hand.

If the decimal part be then multiplied by 4, and five places of decimals pointed off from the right hand, the figures to the left hand will express the roods.

If the decimal part of this result be now multiplied by 40, and five places for decimals pointed off, as before, the figures to the left will express the perches.

If one of the dimensions be in links, and the other in chains, the chains may be reduced to links by annexing two ciphers- or, the multiplication may be made without annexing the ciphers, and the product reduced to acres and decimals of an acre, by pointing off three decimal places at the right hand.

When both dimensions are in chains, the product is reduced to acres by dividing by 10, or pointing off one decimal place.

From which we conclude: that,

I. *If links be multiplied by links, the product is reduced to acres by pointing off five decimal places from the right hand.*

II. *If chains be multiplied by links, the product is reduced to acres by pointing off three decimal places from the right hand.*

III. *If chains be multiplied by chains, the product is reduced to acres by pointing off one decimal place from the right hand.*

7. Since there are 16,5 feet in a rod, a square rod is equal to

$$16{,}5\times 16{,}5=272{,}25 \text{ square feet.}$$

If the last number be multiplied by 160, we shall have

$$272{,}25\times 160=43560 \text{ the square feet in an acre.}$$

Since there are 9 square feet in a square yard, if the last number be divided by 9, we obtain

$$4840=\text{the number of square yards in an acre.}$$

PROBLEM I.

To find the area of a square, a rectangle, a rhombus, or a parallelogram.

RULE.

Multiply the base by the perpendicular height and the product will be the area (Bk. IV. Th. viii).

EXAMPLES.

1. Required the area of the square $ABCD$, each of whose sides is 36 feet.

We multiply two sides of the square together, and the product is the area in square feet.

Operation.

$36 \times 36 = 1296$ *sq. ft.*

2. How many acres, roods, and perches, in a square whose side is 35,25 chains? *Ans.* 124 *A.* 1 *R.* 1 *P.*

3. What is the area of a square whose side is 8 feet 4 inches? *Ans.* 69 *ft.* 5′ 4″.

4. What is the contents of a square field whose side is 46 rods? *Ans.* 13 *A.* 0 *R.* 36 *P.*

5. What is the area of a square whose side is 4769 yards? *Ans.* 22743361 *sq. yds.*

6. What is the area of the parallelogram *ABCD*, of which the base *AB* is 64 feet, and altitude *DE*, 36 feet?

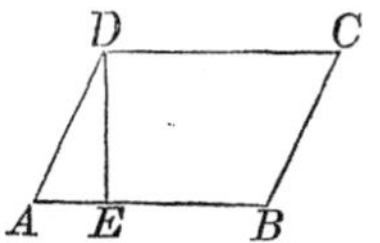

We multiply the base 64, by the perpendicular height 36, and the product is the required area.

Operation.

$64 \times 36 = 2304$ *sq. ft.*

7. What is the area of a parallelogram whose base is 12,25 yards, and altitude 8,5? *Ans.* 104,125 *sq. yds.*

8. What is the area of a parallelogram whose base is 8,75 chains, and altitude 6 chains? *Ans.* 5 *A.* 1 *R.* 0 *P.*

9. What is the area of a parallelogram whose base is 7 feet 9 inches, and altitude 3 feet 6 inches?

Ans. 27 *sq. ft.* 1′ 6″.

10. To find the area of a rectangle $ABCD$, of which the base $AB=45$ yards, and the altitude $AD=15$ yards.

Here we simply multiply the base by the altitude, and the product is the area.

Operation.

$45\times15=675$ *sq. yds.*

11. What is the area of a rectangle whose base is 14 feet 6 inches, and breadth 4 feet 9 inches?

Ans. 68 *sq. ft.* 10′ 6″.

12. Find the area of a rectangular board whose length is 112 feet, and breadth 9 inches. *Ans.* 84 *sq. ft.*

13. Required the area of a rhombus whose base is 10,51 and breadth 4,28 chains. *Ans.* 4 *A.* 1 *R.* 39,7 *P*+.

14. Required the area of a rectangle whose base is 12 feet 6 inches, and altitude 9 feet 3 inches.

Ans. 115 *sq. ft.* 7′ 6″

PROBLEM II.

To find the area of a triangle, when the base and altitude are known.

RULE.

I. *Multiply the base by the altitude, and half the product will be the area.*

II. *Multiply the base by half the altitude and the product will be the area* (Bk. IV. Th. ix).

EXAMPLES.

1. Required the area of the triangle ABC, whose base AB is 10,75 feet, and altitude 7,25 feet.

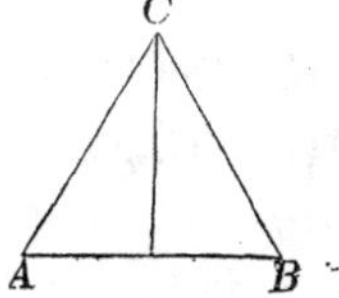

	Operation.
We first multiply the base by the altitude, and then divide the product by 2.	$10{,}75 \times 7{,}25 = 77{,}9375$ and $77{,}9375 \div 2 = 38{,}96875$ = area.

2. What is the area of a triangle whose base is 18 feet 4 inches, and altitude 11 feet 10 inches?
Ans. 108 *sq. ft.* 5′ 8″.

3. What is the area of a triangle whose base is 12,25 chains, and altitude 8,5 chains? *Ans.* 5 *A.* 0 *R.* 33 *P.*

4. What is the area of a triangle whose base is 20 feet, and altitude 10,25 feet. *Ans.* 102,5 *sq. ft.*

5. Find the area of a triangle whose base is 625 and altitude 520 feet. *Ans.* 162500 *sq. ft.*

6. Find the number of square yards in a triangle whose base is 40 and altitude 30 feet. *Ans.* $66\frac{2}{3}$ *sq. yds.*

7. What is the area of a triangle whose base is 72,7 yards, and altitude 36,5 yards? *Ans.* 1326,775 *sq. yds.*

PROBLEM III.

To find the area of a triangle when the three sides are known.

RULE,

I. *Add the three sides together and take half their sum.*

II. *From this half sum take each side separately.*

III. *Multiply together the half sum and each of the three remainders, and then extract the square root of the product which will be the required area.*

EXAMPLES.

1. Find the area of a triangle whose sides are 20, 30, and 40 rods.

```
  20        45              45              45
  30        20              30              40
  40        25 1st rem.     15 2d rem.       5 3d rem.
2)90
  45 half sum,
```

Then, to obtain the product, we have

$$45 \times 25 \times 15 \times 5 = 84375;$$

from which we find

$$\text{area} = \sqrt{84375} = 290{,}4737 \text{ perches.}$$

2. How many square yards of plastering are there in a triangle, whose sides are 30, 40, and 50 feet? *Ans.* $66\frac{2}{3}$.

3. The sides of a triangular field are 49 chains, 50,25 chains, and 25,69: what is its area?

Ans. 61 *A.* 1 *R.* 39,68 *P.*

4. What is the area of an isosceles triangle, whose base is 20, and each of the equal sides 15? *Ans.* 111,803.

5. How many acres are there in a triangle whose three sides are 380, 420 and 765 yards. *Ans.* 9 *A.* 0 *R.* 38 *P.*

6. How many square yards in a triangle whose sides are 13, 14, and 15 feet. *Ans.* $9\frac{1}{3}$.

7 What is the area of an equilateral triangle whose side is 25 feet? *Ans.* 270,6329 *sq. ft.*

8. What is the area of a triangle whose sides are 24, 36, and 48 yards? *Ans.* 418,282 *sq. yds.*

PROBLEM IV.

To find the hypothenuse of a right angled triangle when the base and perpendicular are known.

RULE.

I. *Square each of the sides separately.*

II. *Add the squares together.*

III. *Extract the square root of the sum, which will be the hypothenuse of the triangle* (Bk. IV. Th. xii).

EXAMPLES.

1. In the right angled triangle ABC, we have, $AB=30$ feet, $BC=40$ feet, to find AC.

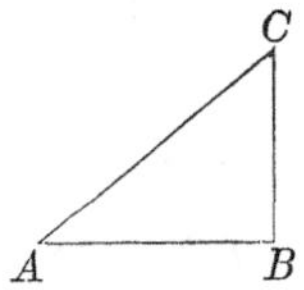

We first square each side, and then take the sum, of which we extract the square root, which gives

$AC=\sqrt{2500}=50$ feet.

Operation.

$\overline{30}^2=900$

$\overline{40}^2=1600$

sum $=2500$

2. The wall of a building, on the brink of a river, is 120 feet high, and the breadth of the river 70 yards: what is the length of a line which would reach from the top of the wall to the opposite edge of the river? *Ans.* 241,86 *ft.*

3. The side roofs of a house of which the eaves are of the same height, form a right angle at the top. Now, the length of the rafters on one side is 10 feet, and on the other 14 feet: what is the breadth of the house? *Ans.* 17,204 *ft.*

4. What would be the width of the house, in the last example, if the rafters on each side were 10 feet?

Ans. 14,142 *ft.*

5. What would be the width, if the rafters on each side were 14 feet? *Ans.* 19,7989 *ft.*

PROBLEM V.

When the hypothenuse and one side of a right angled triangle are known, to find the other side.

RULE.

Square the hypothenuse and also the other given side, and take their difference: extract the square root of this difference, and the result will be the required side (Bk. IV. Th. xii. Cor.).

EXAMPLES.

1. In the right angled triangle ABC, there are given

$AC=50$ feet, and $AB=40$ feet, required the side BC.

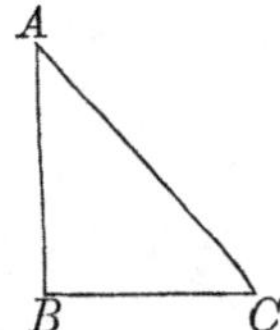

We first square the hypothenuse and the other side, after which we take the difference, and then extract the square root, which gives

Operation.

$\overline{50}^2=2500$
$\overline{40}^2=1600$
Diff.$=$ 900

$$BC=\sqrt{900}=30 \text{ feet.}$$

2. The height of a precipice on the brink of a river is 103 feet, and a line of 320 feet in length will just reach from the top of it to the opposite bank: required the breadth of th river. *Ans.* 302,9703 *ft.*

3. The hypothenuse of a triangle is 53 yards, and the perpendicular 45 yards: what is the base? *Ans.* 28 *yds.*

4. A ladder 60 feet in length, will reach to a window 40

15*

feet from the ground on one side of the street, and by turning it over to the other side, it will reach a window 50 feet from the ground: required the breadth of the street.

Ans. 77,8875 *ft.*

PROBLEM VI.

To find the area of a trapezoid.

RULE.

Multiply the sum of the parallel sides by the perpendicular distance between them, and then divide the product by two: the quotient will be the area (Bk. IV. Th. x).

EXAMPLES.

1. Required the area of the trapezoid *ABCD*, having given

$AB=321,51$ feet, $DC=214,24$ feet, and $CE=171,16$ feet.

We first find the sum of the sides, and then multiply it by the perpendicular height, after which, we divide the product by 2, for the area.

Operation.

$321,51+214,24=535,75=$ sum of parallel sides.

Then,

$535,75\times 171,16=91698,97$

and, $\frac{91698,97}{2}=45849,485$

$=$the area.

2. What is the area of a trapezoid, the parallel sides of which, are 12,41 and 8,22 chains, and the perpendicular distance between them 5,15 chains?

Ans. 5 *A.* 1 *R.* 9,956 *P.*

3. Required the area of a trapezoid whose parallel sides

are 25 feet 6 inches, and 18 feet 9 inches, and the perpendicular distance between them 10 feet and 5 inches.

Ans. 230 *sq. ft.* 5′ 7″.

4. Required the area of a trapezoid whose parallel sides are 20,5 and 12,25, and the perpendicular distance between them 10,75 yards. *Ans.* 176,03125 *sq. yds.*

5. What is the area of a trapezoid whose parallel sides are 7,50 chains, and 12,25 chains, and the perpendicular height 15,40 chains? *Ans.* 15 *A.* 0 *R.* 33,2 *P.*

PROBLEM VII.

To find the area of a quadrilateral.

RULE.

Measure the four sides of the quadrilateral, and also one of the diagonals: the quadrilateral will thus be divided into two triangles, in both of which all the sides will be known. Then, find the areas of the triangles separately, and their sum will be the area of the quadrilateral.

EXAMPLES.

1. Suppose that we have measured the sides and diagonal AC, of the quadrilateral $ABCD$, and found

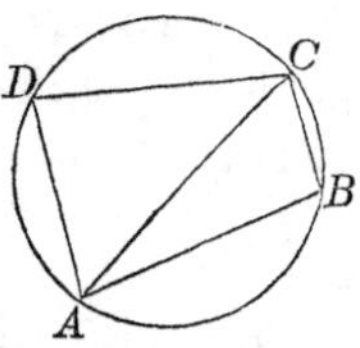

$AB=40{,}05$ chains; $CD=29{,}87$ chains,
$BC=26{,}27$ chains, $AD=37{,}07$ chains,
and $AC=55$ chains:
required the area of the quadrilateral.

Ans. 101 *A.* 1 *R.* 15 *P.*

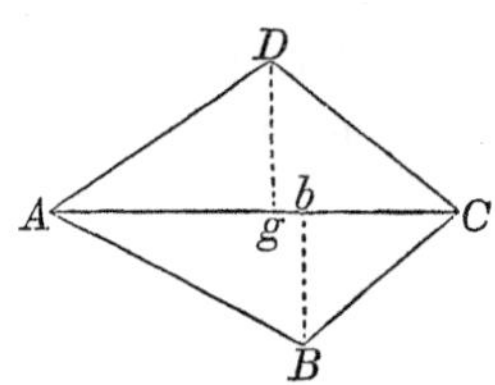

REMARK.—Instead of measuring the four sides of the quadrilateral, we may let fall the perpendicuars Bb, Dg, on the diagonal AC. The area of the triangles may then be determined by measuring these perpendiculars and diagonal AC. The pendiculars are, $Dg=$ 18,95 chains, and $Bb=17,92$ chains.

2. Required the area of a quadrilateral whose diagonal is 80,5, and two perpendiculars 24,5, and 30,1 feet.

Ans. 2197,65 *sq. ft.*

3. What is the area of a quadrilateral whose diagonal is 108 feet 6 inches, and the perpendiculars 56 feet 3 inches, and 60 feet 9 inches? *Ans.* 6347 *sq. ft.* 3′.

4. How many square yards of paving in a quadrilateral whose diagonal is 65 feet, and the two perpendiculars 28, and $33\frac{1}{2}$ feet? *Ans.* $222\frac{1}{12}$ *sq. yds.*

5. Required the area of a quadrilateral whose diagonal is 42 feet, and the two perpendiculars 18, and 16 feet.

Ans. 714 *sq. ft.*

6. What is the area of a quadrilateral in which the diagonal is 320,75 chains, and the two perpendiculars 69,73 chains, and 130,27 chains? *Ans.* 3207 *A.* 2 *R.*

PROBLEM VIII.

To find the area of a regular polygon.

RULE.

Multiply half the perimeter of the figure by the perpendicular let fall from the centre on one of the sides, and the product will be the area (Bk. IV. Th. xxvi).

EXAMPLES.

1. Required the area of the regular pentagon *ABCDE*, each of whose sides *AB*, *BC*, &c., is 25 feet, and the perpendicular *OP*, 17,2 feet.

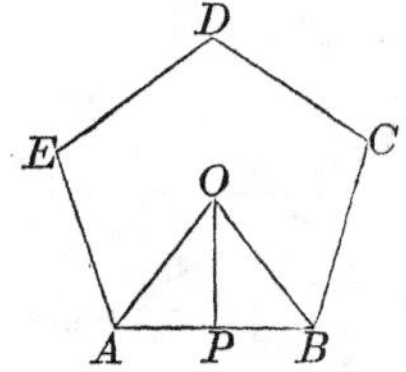

We first multiply one side by the number of sides and divide the product by 2: this gives half the perimeter which we multiply by the perpendicular for the area.

Operation.

$\frac{25\times5}{2}=62,5=$half the perimeter. Then,

$62,5\times17,2=1075$ *sq. ft.*=the area.

2. The side of a regular pentagon is 20 yards, and the perpendicular from the centre on one of the sides 13,76382; required the area.

Ans. 688,191 *sq. yds.*

3. The side of a regular hexagon is 14, and the perpendicular from the centre on one of the sides 12,1243556: required the area.

Ans. 509,2229352 *sq. ft.*

4. Required the area of a regular hexagon whose side is 14 6, and perpendicular from the centre 12,64 feet.

Ans. 553,632 *sq. ft.*

5. Required the area of a heptagon whose side is 19,38 and perpendicular 20 feet.

Ans. 1356,6 *sq. ft.*

The following table shows the areas of the ten regular

polygons when the side of each is equal to 1 : it also shows the length of the radius of the inscribed circle.

Number of sides.	Names.	Areas.	Radius of inscribed circle.
3	Triangle,	0,4330127	0,2886751
4	Square,	1,0000000	0,5000000
5	Pentagon,	1,7204774	0,6881910
6	Hexagon,	2,5980762	0,8660254
7	Heptagon,	3,6339124	1,0382617
8	Octagon,	4,8284271	1,2071068
9	Nonagon,	6,1818242	1,3737387
10	Decagon,	7,6942088	1,5388418
11	Undecagon,	9,3656404	1.2028437
12	Dodecagon,	11,1961524	1,8660254

Now, since the areas of similar polygons are to each other as the squares described on their homologous sides (Bk. IV. Th. xx), we have

$$1^2 \; : \; \text{tabular area} \; :: \; \text{any side squared} \; : \; \text{area}.$$

Hence, to find the area of a regular polygon, we have the following

RULE.

I. *Square the side of the polygon.*

II. *Multiply the square so found, by the tabular area set opposite the polygon of the same number of sides, and the product will be the area.*

EXAMPLES.

1. What is the area of a regular hexagon whose side is 20?

$$\overline{20}^2 = 400 \qquad \text{and tabular area} = 2{,}5980762.$$

Hence,

$$2{,}5980762 \times 400 = 1039{,}23048 = \text{the area}.$$

2. What is the area of a pentagon whose side is 25?
Ans. 1075,298375.

3. What is the area of a heptagon whose side is 30 feet?
Ans. 3270,52116.

4. What is the area of an octagon whose side is 10 feet?
Ans. 482,84271 *sq. ft.*

5. The side of a nonagon is 50: what is its area?
Ans. 15454,5605.

6. The side of an undecagon is 20: what is its area?
Ans. 3746,25616.

7. The side of a dodecagon is 40: what is its area?
Ans. 17913,84384

PROBLEM IX.

To find the area of a long and irregular figure, bounded on one side by a straight line.

RULE.

I. *Divide the right line or base into any number of equal parts, and measure the breadth of the figure at the points of division, and also at the extremities of the base.*

II. *Add together the intermediate breadths, and half the sum of the extreme ones.*

III. *Multiply this sum by the base line, and divide the product by the number of equal parts of the base.*

EXAMPLES.

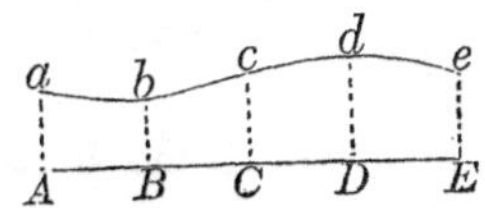

1. The breadths of an irregular figure, at five equidistant places, A, B, C, D, and E, being 8,20 chains, 7,40 chains,

9,20 chains, 10,20 chains, and 8,60 chains, and the whole length 40 chains: required the area.

```
     8,20                         35,20
     8,60                            40
  2)16,80                     4)1408,00
     8,40 mean of the extremes.  352,00 square chains.
     7,40
     9,20
    10,20
    35,20 the sum.
```

Ans. 35 *A.* 32 *P.*

2. The length of an irregular piece of land being 21 chains and the breadths, at six equidistant points, being 4,35 chains, 5,15 chains, 3,55 chains, 4,12 chains, 5,02 chains, and 6,10 chains: required the area. *Ans.* 9 *A.* 2 *R.* 30 *P.*

3. The length of an irregular figure is 84 yards, and the breadths at six equidistant places are 17,4; 20,6; 14,2; 16,5; 20,1; and 24,4: what is the area? *Ans.* 1550,64 *sq. yds.*

4. The length of an irregular field is 39 rods, and its breadths at five equidistant places, are 4,8; 5,2; 4,1; 7,3, and 7,2 rods: what is its area? *Ans.* 220,35 *sq. rods.*

5. The length of an irregular field is 50 yards, and its breadths at seven equidistant points, are 5,5; 6,2; 7,3; 6; 7,5; 7; and 8,8 yards: what is its area?

Ans. 342,916 *sq. yds.*

6. The length of an irregular figure being 37,6, and the breadths at nine equidistant places, 0; 4,4; 6,5; 7,6; 5,4; 8; 5,2; 6,5; and 6,1: what is the area? *Ans.* 219,255.

PROBLEM X.

To find the circumference of a circle when the diameter is known.

RULE

Multiply the diameter by 3,1416, *and the product will be the circumference.*

EXAMPLES.

1. What is the circumference of a circle whose diameter is 17?

We simply multiply the number 3,1416 by the diameter, and the product is the circumference.

Operation.

3,1416×17=53,4072,

which is the circumference.

2. What is the circumference of a circle whose diameter is 40 feet? *Ans.* 125,664 *ft.*

3. What is the circumference of a circle whose diameter is 12 feet? *Ans.* 37,6992 *ft.*

4. What is the circumference of a circle whose diameter is 22 yards? *Ans.* 69,1152 *yds.*

5. What is the circumference of the earth—the mean diameter being about 7921 miles? *Ans.* 24884,6136 *mi.*

PROBLEM XI.

To find the diameter of a circle when the circumference is known.

RULE.

Divide the circumference by the number 3,1416, *and the quotient will be the diameter.*

EXAMPLES.

1. The circumference of a circle is 69,1152 yards: what is the diameter?

We simply divide the circumference by 3,1416, and the quotient 22 is the diameter sought.	*Operation.* 3,1416)69,1152(22 62832 62832 62832

2. What is the diameter of a circle whose circumference is 11652,1944 feet? *Ans.* 3709.

3. What is the diameter of a circle whose circumference is 6850? *Ans.* 2180,4176.

4. What is the diameter of a circle whose circumference is 50? *Ans.* 15,915.

5. If the circumference of a circle is 25000,8528, what is the diameter? *Ans.* 7958.

PROBLEM XII.

To find the length of a circular arc, when the number of degrees which it contains, and the radius of the circle are known.

RULE.

Multiply the number of degrees by the decimal ,01745, *and the product arising by the radius of the circle.*

EXAMPLES.

1. What is the length of an arc of 30 degrees, in a circle whose radius is 9 feet.

We merely multiply the given decimal by the number of degrees, and by the radius.	*Operation.* $,01745\times30\times9=4,7115$, which is the length of the arc.

REMARK.—When the arc contains degrees and minutes, reduce the minutes to the decimals of a degree, which is done by dividing them by 60.

2. What is the length of an arc containing 12° 10′ or $12\frac{1}{6}$,° the diameter of the circle being 20 yards?

Ans. 2,1231.

3. What is the length of an arc of 10° 15′ or $10\frac{1}{4}$°, in circle whose diameter is 68? *Ans.* 6,0813.

PROBLEM XIII.

To find the length of the arc of a circle when the chord and radius are given.

RULE.

I. *Find the chord of half the arc.*

II. *From eight times the chord of half the arc, subtract the chord of the whole arc, and divide the remainder by* 3, *and the quotient will be the length of the arc, nearly.*

EXAMPLES.

1. The chord $AB=30$ feet, and the radius $AC=20$ feet: what is the length of the arc ADB?

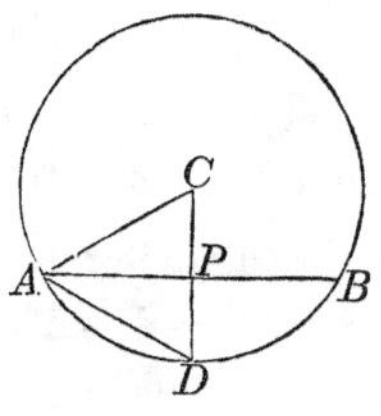

First draw CD perpendicular to the chord AB: it will bisect the chord at P, and the arc of the chord at D. Then $AP=15$ feet. Hence,

$$\overline{AC}^2-\overline{AP}^2=\overline{CP}^2: \text{ that is,}$$

$$400-225=175 \quad \text{and} \quad \sqrt{175}=13,228=CP.$$

Then $\quad CD-CP=20-13,228=6,772=DP.$

Again, $\quad AD=\sqrt{\overline{AP}^2+\overline{PD}^2}=\sqrt{225+45,859984}:$

hence, $\quad AD=16,4578=$chord of the half arc.

Then, $\quad \dfrac{16,4578\times 8-30}{3}=33,8874=\text{arc } ADB.$

2. What is the length of an arc the chord of which is 24 feet, and the radius of the circle 20 feet?

Ans. 25,7309 *ft.*

3. The chord of an arc is 16 and the diameter of the circle 20: what is the length of the arc? *Ans.* 18,5178.

4. The chord of an arc is 50, and the chord of half the arc is 27: what is the length of the arc? *Ans.* $55\frac{1}{3}$.

PROBLEM XIV.

To find the area of a circle when the diameter and circumference are both known.

RULE.

Multiply the circumference by half the radius and the product will be the area (Bk. IV. Th. xxvii).

EXAMPLES.

1. What is the area of a circle whose diameter is 10, and circumference 31,416?

If the diameter be 10, the radius is 5, and half the radius is $2\frac{1}{2}$: hence, the circumference multiplied by $2\frac{1}{2}$ gives the area.

Operation.

$31,416 \times 2\frac{1}{2} = 78,54$; which is the area.

2. Find the area of a circle whose diameter is 7; and circumference 21,9912 yards. *Ans.* 38,4846 *yds.*

3. How many square yards in a circle whose diameter is $3\frac{1}{2}$ feet, and circumference 10,9956. *Ans.* 1,069016.

4. What is the area of a circle whose diameter is 100, and circumference 314,16? *Ans.* 7854.

5. What is the area of a circle whose diameter is 1, and circumference 3,1416. *Ans.* 0,7854.

6. What is the area of a circle whose diameter is 40, and circumference 131,9472? *Ans.* 1319,472.

PROBLEM XV.

To find the area of a circle when the diameter only is known.

RULE.

Square the diameter, and then multiply by the decimal ,7854.

EXAMPLES.

What is the area of a circle whose diameter is 5?

We square the diameter, which gives us 25, and we then multiply this number and the decimal ,7854 together.

Operation.

$$\begin{array}{r} ,7854 \\ \overline{5}^2 = \quad 25 \\ \hline 39270 \\ 15708 \\ \hline \text{area} = 19,6350 \end{array}$$

2. What is the area of a circle whose diameter is 7? *Ans.* 38,4846.

3. What is the area of a circle whose diameter is 4,5? *Ans.* 15,90435.

4. What is the number of square yards in a circle whose diameter is $1\frac{1}{6}$ yards? *Ans.* 1,069016.

5. What is the area of a circle whose diameter is 8,75 feet? *Ans.* 60,1322 *sq. ft.*

PROBLEM XVI.

To find the area of a circle when the circumference onlv is known.

RULE.

Multiply the square of the circumference by the decimal ,07958, and the product will be the area very nearly.

EXAMPLES.

1. What is the area of a circle whose circumference is 3,1416?

We first square the circumference, and then multiply by the decimal ,07958.

Operation.

$$\overline{3{,}1416}^2 = 9{,}86965056$$
$$,07958$$
$$\text{area} = \overline{,7854+}$$

2. What is the area of a circle whose circumference is 91?
Ans. 659,00198.

3. Suppose a wheel turns twice in tracking $16\frac{1}{2}$ feet, and that it turns just 200 times in going round a circular bowling-green: what is the area in acres, roods, and perches?
Ans. 4 *A.* 3 *R.* 35,8 *P.*

4. How many square feet are there in a circle whose circumference is 10,9956 yards? *Ans.* 86,5933.

5. How many perches are there in a circle whose circumference is 7 miles? *Ans.* 399300,608.

PROBLEM XVII.

Having given a circle, to find a square which shall have an equal area.

RULE.

I. *The diameter* $\times$,8862 = *side of an equivalent square.*

II. *The circumference* $\times$,2821 = *side of an equivalent square*

EXAMPLES.

1. The diameter of a circle is 100: what is the side of a square of equal area? *Ans.* 88,62.

2. The diameter of a circular fishpond is 20 feet, what would be the side of a square fishpond of an equal area? *Ans.* 17,724 *ft.*

3. A man has a circular meadow of which the diameter is 875 yards, and wishes to exchange it for a square one of equal size: what must be the side of the square? *Ans.* 775,425.

4. The circumference of a circle is 200: what is the side of a square of an equal area? *Ans.* 56,42.

5. The circumference of a round fishpond is 400 yards: what is the side of a square pond of equal area? *Ans.* 112,84.

6. The circumference of a circular bowling-green is 412 yards: what is the side of a square one of equal area? *Ans.* 116,2252 *yds.*

7. The circumference of a circular walk is 625: what is the side of a square containing the same area? *Ans.* 176,3125.

PROBLEM XVIII.

Having given the diameter or circumference of a circle, to find the side of the inscribed square.

RULE.

I. *The diameter* × ,7071 = *side of the inscribed square.*

II. *The circumference* × ,2251 = *side of the inscribed square*

EXAMPLES.

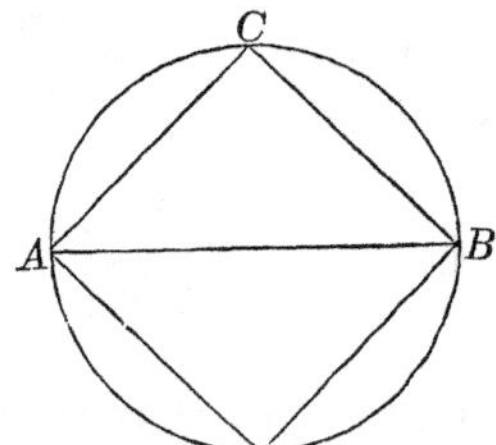

1. The diameter AB of a circle is 400: what is the value of AC, the side of the inscribed square?

Here,

$,7071 \times 400 = 282,8400 = AC.$

2. The diameter of a circle is 412 feet: what is the side of the inscribed square? *Ans.* 291,3252 *ft.*

3. If the diameter of a circle be 600 what is the side of the inscribed square? *Ans.* 424,26.

4. The circumference of a circle is 312 feet: what is the side of the inscribed square? *Ans.* 70,2312 *ft.*

5. The circumference of a circle is 819 yards: what is the side of the inscribed square? *Ans.* 184,3569 *yds.*

6. The circumference of a circle is 715: what is the side of the inscribed square? *Ans.* 160,9465.

7. The circumference of a circular walk is 625: what is the side of an inscribed square? *Ans.* 140,6875.

PROBLEM XIX.

To find the area of a circular sector.

RULE.

I. *Find the length of the arc by Problem* XII.

II. *Multiply the arc by one half the radius, and the product will be the area*

EXAMPLES.

1. What is the area of the circular sector ACB, the arc AB containing 18°, and the radius CA being equal to 3 feet.

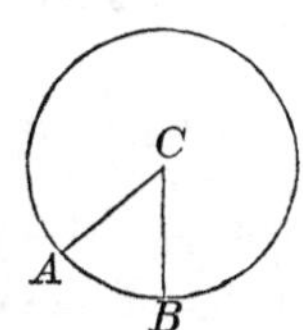

First, ,01745 × 18 × 3 = ,94230 = length AB.
Then, ,94230 × $1\frac{1}{2}$ = 1,41345 = area.

2. What is the area of a sector of a circle in which the radius is 20 and the arc one of 22 degrees?

Ans. 76,7800.

3. Required the area of a sector whose radius is 25 and the arc of 147° 29′. *Ans.* 804,2448.

4. Required the area of a semicircle in which the radius is 13. *Ans.* 265,4143.

5. What is the area of a circular sector when the length of the arc is 650 feet and the radius 325?

Ans. 105625 *sq. ft*

PROBLEM XX.

To find the area of a segment of a circle.

RULE.

I. *Find the area of the sector having the same arc with the segment, by the last Problem.*

II. *Find the area of the triangle formed by the chord of the segment and the two radii through its extremities.*

III. *If the segment is greater than the semicircle, add the two areas together; but if it is less, subtract them, and the result in either case, will be the area required.*

EXAMPLES.

1. What is the area of the segment ADB, the chord $AB=24$ feet and $CA=20$ feet.

First, $CP=\sqrt{\overline{CA}^2-\overline{AP}^2}$
$=\sqrt{400-144}=16$

Then,
$PD=CD-CP=20-16=4.$

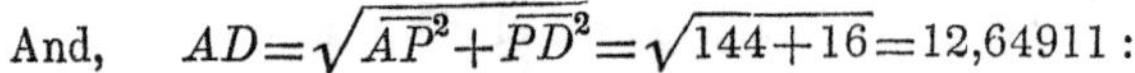

And, $AD=\sqrt{\overline{AP}^2+\overline{PD}^2}=\sqrt{144+16}=12,64911$:

then, arc $ADB=\dfrac{12,64911\times8-24}{3}=25,7309.$

Arc	$ADB=25,7309$	$AP=12$
half radius	$=\quad 10$	$CP=16$
area sector	$ADBC=257,3090$	area $CAB=192$
area	$CAB=192$	
	$65,309=$area of segment ADB	

2. Find the area of the segment AFB, knowing the following lines, viz: $AB=20,5$; $FP=17,17$; $AF=20$; $FG=11,5$; and $CA=11,64$.

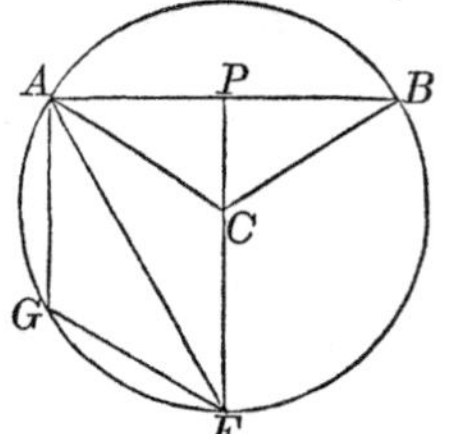

$$\text{Arc } AGF=\frac{FG\times8-AF}{3}=\frac{11,5\times8-20}{3}=24:$$

and sector $AGFBC=24\times11,64=279,36$:

but $CP=FP-AC=17,17-11,64=5,53$:

Then, area $ACB=\dfrac{AB\times CP}{2}=\dfrac{20,5\times5,53}{2}=56,6825.$

Then, area of sector $AFBC$=279,36
do. of triangle ABC= 56,6825
gives area of segment AFB=336,0425

3 What is the area of a segment; the radius of the circle being 10, and the chord of the arc 12 yards?

Ans. 16,324 *sq. yds.*

4. Required the area of the segment of a circle whose chord is 16, and the diameter of the circle 20.

Ans. 44,5903.

5. What is the area of a segment whose arc is a quadrant, the diameter of the circle being 18? *Ans.* 63,6174.

6. The diameter of a circle is 100, and the chord of the segment 60: what is the area of the segment?

Ans. 408, nearly

PROBLEM XXI.

To find the area of an ellipse.

Multiply the two axes together, and their product by the decimal ,7854, *and the result will be the required area.*

EXAMPLES.

1. Required the area of an ellipse, whose transverse axis AB=70 feet, and the conjugate axis DE=50 feet.

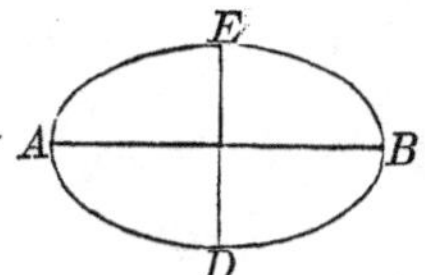

$AB \times DE = 70 \times 50 = 3500$:

Then, ,7854 × 3500=2748,9=area.

2. Required the area of an ellipse whose axes are 24 and 18. *Ans.* 339,2928.

3. What is the area of an ellipse whose axes are 80 and 60? *Ans.* 3769,92.

4. What is the area of an ellipse whose axes are 50 and 45? *Ans.* 1767,15.

PROBLEM XXII.

To find the area of a circular ring: that is, the area included between the circumferences of two circles, having a common centre.

RULE.

I. *Square the diameter of each ring, and subtract the square of the less from that of the greater.*

II. *Multiply the difference of the squares by the decimal* ,7854, *and the product will be the area.*

EXAMPLES.

1. In the concentric circles having the common centre C, we have

$AB=10$ yds., and $DE=6$ yards: what is the area of the space included between them?

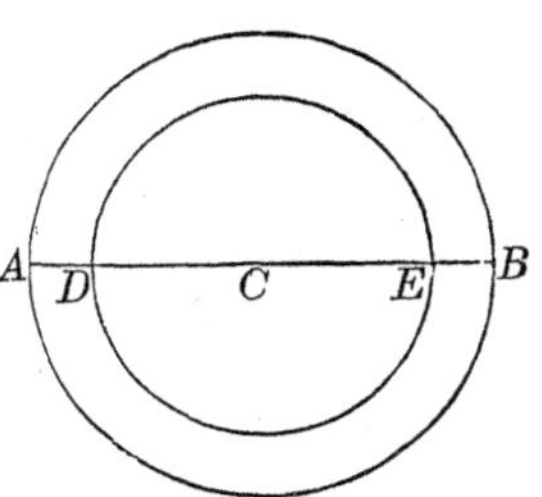

$$\overline{BA}^2=\overline{10}^2=100$$
$$\overline{DE}^2=\overline{6}^2=36$$
$$\text{Difference}=64$$

Then, $63\times,7854=50,2656=$area.

2. What is the area of the ring when the diameters of the circle are 20 and 10? *Ans.* 235,62.

3. If the diameters are 20 and 15, what will be the area included between the circumferences? *Ans.* 137,445.

4. If the diameters are 16 and 10, what will be the area included between the circumferences? *Ans.* 122,5224.

5. Two diameters are 21,75 and 9,5; required the area of the circular ring. *Ans.* 300,6609.

6. If the two diameters are 4 and 6, what is the area of the ring? *Ans.* 15,708.

MENSURATION OF SOLIDS.

DEFINITIONS.

The mensuration of solids is divided into two parts.

1st, The mensuration of the surfaces of solids: and

2d, The mensuration of their solidities.

We have already seen that the unit of measure for plane surfaces, is a square whose side is the unit of length (Bk. IV. Def. 7).

2. A curve line which is expressed by numbers is also referred to an unit of length, and its numerical value is the number of times which the line contains the unit.

If then, we suppose the linear unit to be reduced to a straight line, and a square constructed on this line, this square will be the unit of measure for curved surfaces.

3. The unit of solidity is a cube, whose edge is the unit in which the linear dimensions of the solid are expressed; and

the face of this cube is the superficial unit in which the surtace of the solid is estimated (Bk. VI. Th. xiii. Sch).

4. The following is a table of solid measure.

1 cubic foot	$=1728$	cubic inches.
1 cubic yard	$=27$	cubic feet.
1 cubic rod	$=4492\frac{1}{8}$	cubic feet.
1 ale gallon	$=282$	cubic inches.
1 wine gallon	$=231$	cubic inches.
1 bushel	$=2150{,}42$	cubic inches.

PROBLEM I.

To find the surface of a right prism.

RULE.

Multiply the perimeter of the base by the altitude and the product will be the convex surface: and to this add the area of the bases, when the entire surface is required (Bk. VI. Th. i).

EXAMPLES.

1. Find the entire surface of the regular prism whose base is the regular polygon $ABCDE$ and altitude AF, when each side of the base is 20 feet and the altitude AF, 50 feet.

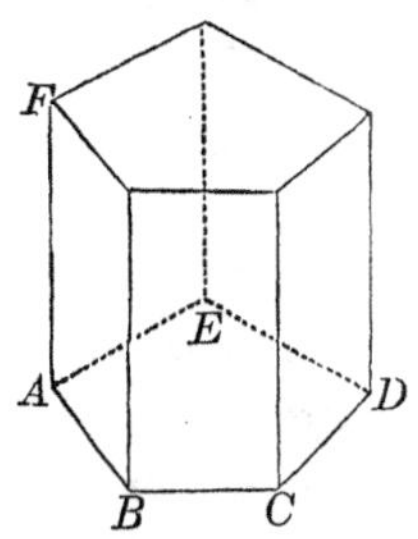

$AB+BC+CD+DE+EA=100$; and $AF=50$: then

$$(AB+BC+CD+DE+EA)\times AF=\text{convex surface}$$

which becomes, $100 \times 50 = 5000$ square feet; which is the convex surface. For the area of the end, we have

$$\overline{AB}^2 \times \text{tabular number} = \text{area } ABCDE,$$

that is, $\overline{20}^2 \times$ tabular number, or $400 \times 1{,}720477 = 688{,}1908 =$ the area $ABCDE$.

Then, convex surface	5000	square feet.
lower base	688,1908	square feet.
upper base	688,1908	square feet.
Entire surface	6376,3816	

2. What is the surface of a cube, the length of each side being 20 feet? *Ans.* 2400 *sq. ft.*

3. Find the entire surface of a triangular prism, whose base is an equilateral triangle, having each of its sides equal to 18 inches, and altitude 20 feet. *Ans.* 91,949 *sq. ft.*

4. What is the convex surface of a regular octagonal prism, the side of whose base is 15 and altitude 12 feet?

Ans. 1440 *sq. ft.*

5. What must be paid for lining a rectangular cistern with lead at 2*d* a pound, the thickness of the lead being such as to require 7*lb.* for each square foot of surface; the inner dimensions of the cistern being as follows: viz. the length 3 feet 2 inches, the breadth 2 feet 8 inches, and the depth 2 feet 6 inches? *Ans.* £2 3*s.* $10\frac{5}{9}d$.

PROBLEM II.

To find the solidity of a prism.

RULE.

Multiply the area of the base by the perpendicular height, and the product will be the solidity.

EXAMPLES.

1. What is the solidity of a regular pentagonal prism whose altitude is 20, and each side of the base 15 feet?

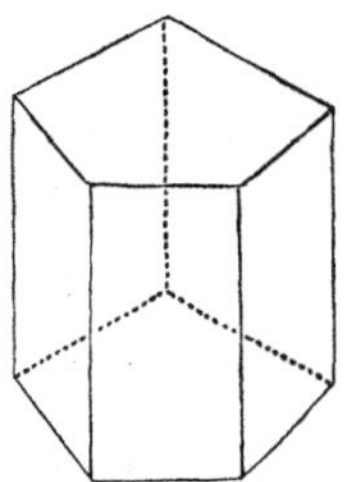

To find the area of the base we have by Problem VIII. page 178.

$\overline{15}^2=225$: and $225\times1,7204774=387,107415=$ the area of the base: hence,

$$387,107415\times20=7742,1483=\text{solidity}.$$

2. What is the solid contents of a cube whose side is 24 inches? *Ans.* 13824 *solid in.*

3. How many cubic feet in a block of marble, of which the length is 3 feet 2 inches, breadth 2 feet 8 inches, and height or thickness 2 feet 6 inches? *Ans.* $21\frac{1}{9}$ *solid ft.*

4. How many gallons of water, ale measure, will a cistern contain whose dimensions are the same as in the last example? *Ans.* $129\frac{17}{47}$

5. Required the solidity of a triangular prism whose altitude is 10 feet, and the three sides of its triangular base 3, 4, and 5 feet. *Ans.* 60 *solid ft.*

6. What is the solidity of a square prism whose height is $5\frac{1}{2}$ feet, and each side of the base $1\frac{1}{3}$ foot?

Ans. $9\frac{7}{9}$ *solid ft.*

7. What is the solidity of a prism whose base is an equilateral triangle, each side of which is 4 feet, the height of the prism being 10 feet? *Ans.* 69,282 *solid ft.*

8. What is the number of cubic or solid feet in a regular pentagonal prism of which the altitude is 15 feet and each side of the base 3,75 feet? *Ans.* 362,913.

PROBLEM III.

To find the surface of a regular pyramid.

RULE.

Multiply the perimeter of the base by half the slant height, and the product will be the convex surface: to this add the area of the base, if the entire surface is required (Bk. VI. Th. vi).

EXAMPLES.

1. In the regular pentagonal pyramid *S—ABCDE*, the slant height *SF* is equal to 45, and each side of the base is 15 feet: required the convex surface, and also the entire surface.

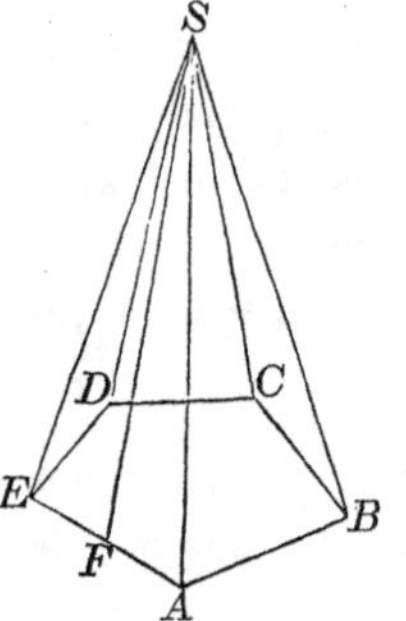

$15\times5=75=$perimeter of the base,
$75\times22\frac{1}{2}=1687,5$ square feet$=$area of convex surface.

And $\overline{15}^2=225$: then $225\times1,7204774=387,107415=$the area of the base.

Hence, convex surface $=1687,5$
area of the base$=$ 387,107415
Entire surface $=$2074,607415 square feet.

2. What is the convex surface of a regular triangular pyramid, the slant height being 20 feet, and each side of the base 3 feet? *Ans.* 90 *sq. ft.*

3. What is the entire surface of a regular pyramid whose slant height is 15 feet, and the base a regular pentagon, of which each side is 25 feet? *Ans.* 2012,798 *sq. ft.*

PROBLEM IV.

To find the convex surface of the frustum of a regular pyramid.

RULE.

Multiply half the sum of the perimeters of the two bases by the slant height of the frustum, and the product will be the convex surface (Bk. VI. Th. vii).

EXAMPLES.

1. In the frustum of the regular pentagonal pyramid each side of the lower base is 30, and each side of the upper base is 20 feet, and the slant height fF is equal to 15 feet. What is the convex surface of the frustum?

Ans. 1875 *sq. ft.*

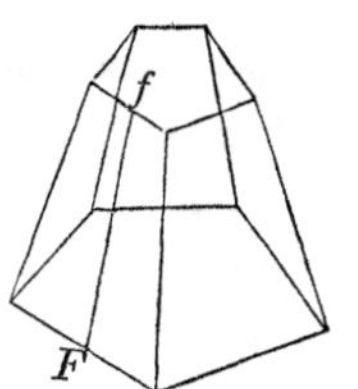

2. How many square feet are there in the convex surface of the frustum of a square pyramid, whose slant height is 10 feet, each side of the lower base 3 feet 4 inches, and each side of the upper base 2 feet 2 inches? *Ans.* 110.

3. What is the convex surface of the frustum of a heptagonal pyramid whose slant height is 55 feet, each side of the lower base 8 feet, and each side of the upper base 4 feet? *Ans.* 2310 *sq. ft.*

PROBLEM V.

To find the solidity of a pyramid.

RULE.

Multiply the area of the base by the altitude and divide the product by 3, the quotient will be the solidity (Bk. VI. Th. xvii)

EXAMPLES.

1 What is the solidity of a pyramid the area of whose base is 215 square feet and the altitude $SO=45$ feet?

First, $215\times45=9675$:

then, $9675\div\ 3=3225$

which is the solidity expressed in solid feet.

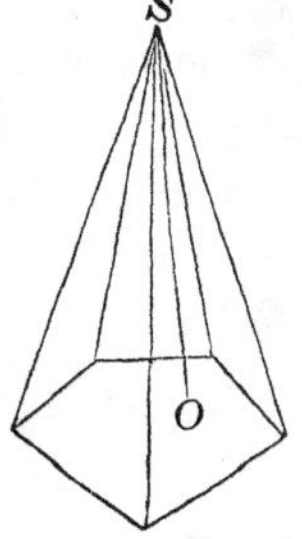

2. Required the solidity of a square pyramid, each side of its base being 30 and its altitude 25. *Ans.* 7500 *solid ft.*

3. How many solid yards are there in a triangular pyramid whose altitude is 90 feet, and each side of its base 3 yards?
Ans. 38,97117.

4. How many solid feet in a triangular pyramid the altitude of which is 14 feet 6 inches, and the three sides of its base 5, 6 and 7 feet? *Ans.* 71,0352.

5. What is the solidity of a regular pentagonal pyramid, its altitude being 12 feet, and each side of its base 2 feet?
Ans. 27,5276 *solid ft.*

6. How many solid feet in a regular hexagonal pyramid, whose altitude is 6,4 feet, and each side of the base 6 inches? *Ans.* 1,38564.

7. How many solid feet are contained in a hexagonal pyramid the height of which is 45 feet, and each side of the base 10 feet? *Ans.* 3897,1143.

8. The spire of a church is an octagonal pyramid, each side of the base being 5 feet 10 inches, and its perpendicular height 45 feet. Within is a cavity, or hollow part, each side of the base being 4 feet 11 inches, and its perpendicular height 41 feet: how many yards of stone does the spire contain? *Ans.* 32,197353

PROBLEM VI.

To find the solidity of the frustum of a pyramid.

RULE.

Add together the areas of the two bases of the frustum and a geometrical mean proportional between them; and then multiply the sum by the altitude, and take one-third the product for the solidity.

EXAMPLES.

1. What is the solidity of the frustum of a pentagonal pyramid the area of the lower base being 16 and of the upper base 9 square feet, the altitude being 7 feet?

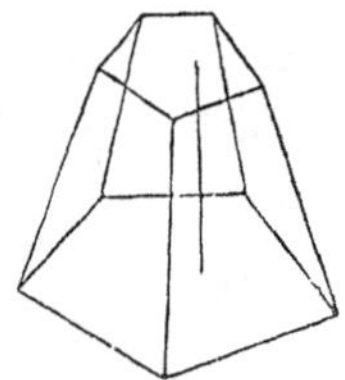

First, $16 \times 9 = 144$: then, $\sqrt{144} = 12$, the mean.

Then, area of lower base $= 16$

area of upper base $= 9$

mean of bases $= 12$

$$\begin{array}{lr} \text{area of lower base} & =16 \\ \text{area of upper base} & =9 \\ \text{mean of bases} & =12 \\ \hline & 37 \\ \text{height} & 7 \\ \hline & 3)\ 259 \\ \hline \text{solidity} & =86\frac{1}{3} \end{array}$$

solidity $= 86\frac{1}{3}$ *solid ft.*

2. What is the number of solid feet in a piece of timber whose bases are squares, each side of the lower base being 15 inches, and each side of the upper base being 6 inches, the length being 24 feet? *Ans.* 19,5.

3. Required the solidity of a regular pentagonal frustum, whose altitude is 5 feet, each side of the lower base 18 inches, and each side of the upper base 6 inches.

Ans. 9,31925 *solid ft.*

4. What is the contents of a regular hexagonal frustum, whose height is 6 feet, the side of the greater end 18 inches, and of the less end 12 inches? *Ans.* 24,681724 *cubic ft.*

5. How many cubic feet in a square piece of timber, the areas of the two ends being 504 and 372 inches, and its length $31\frac{1}{2}$ feet? *Ans.* 95,447.

6. What is the solidity of a squared piece of timber, its length being 18 feet, each side of the greater base 18 inches and each side of the smaller 12 inches?

Ans. 28,5 *cubic ft.*

7. What is the solidity of the frustum of a regular hexagonal pyramid, the side of the greater end being 3 feet, that of the less 2 feet, and the height 12 feet?

Ans. 197,453776 *solid ft.*

MEASURES OF THE THREE ROUND BODIES.

PROBLEM I.

To find the surface of a cylinder.

RULE.

Multiply the circumference of the base by the altitude, and the product will be the convex surface; and to this, add the areas of the two bases, when the entire surface is required (Bk. VI. Th. ii).

EXAMPLES.

1. What is the entire surface of the cylinder in which AB, the diameter of the base, is 12 feet, and the altitude EF 30 feet?

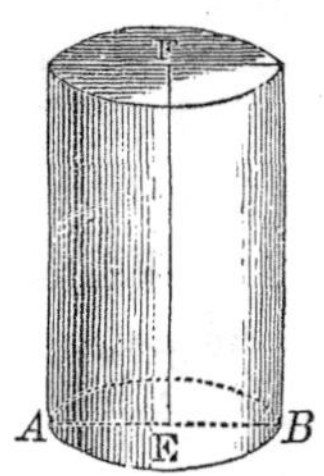

First, to find the circumference of the base, (Prob. X. page 180): we have $3,1416 \times 12 = 37,6992 =$ circumference of the base.

Then, $37,6992 \times 30 = 1130,9760 =$ convex surface.

Also, $\overline{12}^2 = 144$: and $144 \times ,7854 = 113,0976 =$ area of the base.

Then,

convex surface	=	1130,9760
lower base		113,0976
upper base		113,0976
Entire area	=	1357,1712

2. What is the convex surface of a cylinder, the diameter of whose base is 20, and the altitude 50 feet?

Ans. 3141,6 *sq. ft.*

3. Required the entire surface of a cylinder, whose altitude is 20 feet, and the diameter of the base 2 feet.

Ans. 131,9472 *ft.*

4. What is the convex surface of a cylinder, the diameter of whose base is 30 inches, and altitude 5 feet?

Ans. 5654,88 *sq. in.*

5. Required the convex surface of a cylinder, whose altitude is 14 feet, and the circumference of the base 8 feet 4 inches. *Ans.* 116,6666, &c., *sq. ft.*

PROBLEM II.

To find the solidity of a cylinder.

RULE.

Multiply the area of the base by the altitude, and the product will be the solidity.

EXAMPLES.

1. What is the solidity of a cylinder, the diameter of whose base is 40 feet, and altitude EF, 25 feet?

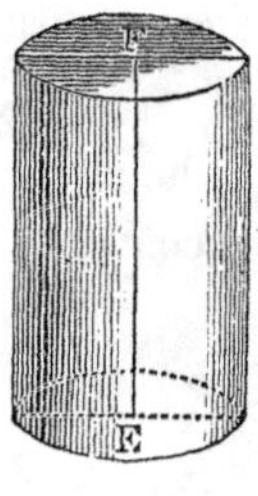

First, to find the area of the base, we have (Prob. xv. page 185),

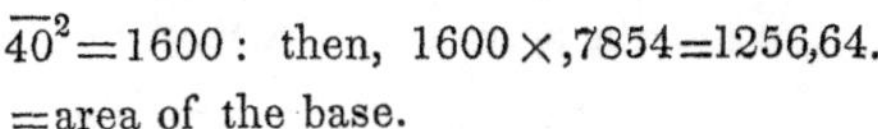

$\overline{40}^2=1600$: then, $1600\times,7854=1256,64$. $=$area of the base.

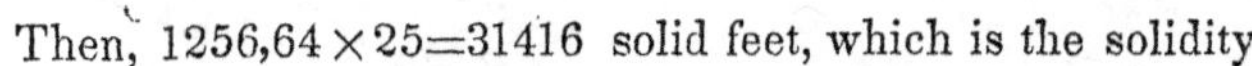

Then, $1256,64\times 25=31416$ solid feet, which is the solidity

2. What is the solidity of a cylinder, the diameter of whose base is 30 feet, and altitude 50 feet?

Ans. 35343 *cubic ft.*

3. What is the solidity of a cylinder whose height is 5 feet, and the diameter of the end 2 feet? *Ans.* 15,708 *solid ft*

4. What is the solidity of a cylinder whose height is 20 feet, and the circumference of the base 20 feet?

Ans. 636,64 *cubic ft.*

5. The circumference of the base of a cylinder is 20 feet, and the altitude 19,318 feet: what is the solidity?

Ans. 614,93 *cubic ft.*

6. What is the solidity of a cylinder whose altitude is 12 feet, and the diameter of its base 15 feet?

Ans. 2120,58 *cubic ft.*

7. Required the solidity of a cylinder whose altitude is 20 feet, and the circumference of whose base is 5 feet 6 inches?

Ans. 48,1459 *cubic ft.*

8. What is the solidity of a cylinder, the circumference of whose base is 38 feet, and altitude 25 feet?

Ans. 2872,838 *cubic ft.*

9. What is the solidity of a cylinder, the circumference of whose base is 40 feet, and altitude 30 feet?

10. The diameter of the base of a cylinder is 84 yards, and the altitude 21 feet: how many solid or cubic yards does it contain? *Ans.* 38792,4768.

PROBLEM III.

To find the surface of a cone.

RULE.

Multiply the circumference of the base by the slant height, and divide the product by 2; the quotient will be the convex surface, to which add the area of the base, when the entire surface is required (Bk. VI. Th. viii).

EXAMPLES.

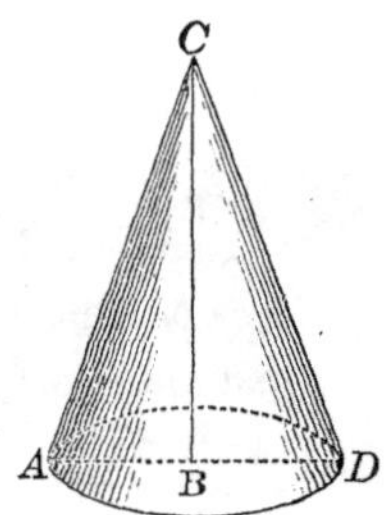

1. What is the convex surface of the cone whose vertex is C, the diameter AD, of its base being $8\frac{1}{2}$ feet, and the side CA, 50 feet.

First, $3{,}1416 \times 8\frac{1}{2} = 26{,}7036 =$ circumference of base.

Then $\dfrac{26{,}7036 \times 50}{2} = 667{,}59 =$ convex surface.

2. Required the entire surface of a cone whose side is 36, and the diameter of its base 18 feet.

Ans. 1272,348 *sq. ft.*

3. The diameter of the base is 3 feet, and the slant height 15 feet: what is the convex surface of the cone?

Ans. 70,686 *sq. ft.*

4. The diameter of the base of a cone is 4,5 feet, and the slant height 20 feet: what is the entire surface?

Ans. 157,27635 *sq. ft.*

5. The circumference of the base of a cone is 10,75, and the slant height is 18,25: what is the entire surface?

Ans. 107,29021 *sq. ft.*

PROBLEM IV.

To find the solidity of a cone.

RULE.

Multiply the area of the base by the altitude; and divide the product by 3, *the quotient will be the solidity* (Bk. VI. Th. xviii).

EXAMPLES.

1. What is the solidity of a cone, the area of whose base is 380 square feet, and altitude CB, 48 feet?

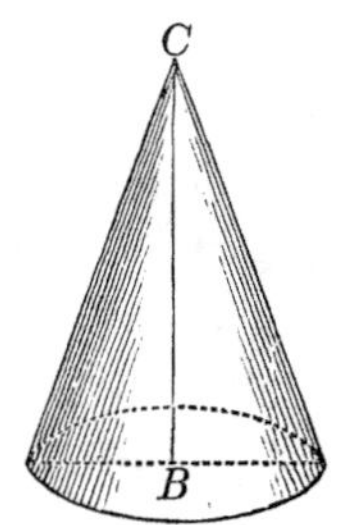

We simply multiply the area of the base by the altitude, and then divide the product by 3.

Operation.

```
      380
       48
     3040
    1520
  3)18240
area=6080
```

2. Required the solidity of a cone whose altitude is 27 feet, and the diameter of the base 10 feet.

Ans. 706,86 *cubic ft.*

3. Required the solidity of a cone whose altitude is $10\frac{1}{2}$ feet, and the circumference of its base 9 feet?

Ans. 22,5609 *cubic ft.*

4. What is the solidity of a cone, the diameter of whose base is 18 inches, and altitude 15 feet?

Ans. 8,83575 *cubic ft.*

5. The circumference of the base of a cone is 40 feet, and the altitude 50 feet: what is the solidity?

Ans. 2122,1333 *solid ft.*

PROBLEM V.

To find the surface of the frustum of a cone.

RULE.

Add together the circumferences of the two bases; and multiply the sum by half the slant height of the frustum; the product will be the convex surface, to which add the areas of the bases, when the entire surface is required (Bk. VI. Th. ix).

EXAMPLES.

1. What is the convex surface of the frustum of a cone, of which the slant height is $12\frac{1}{2}$ feet, and the circumferences of the bases 8,4 and 6 feet.

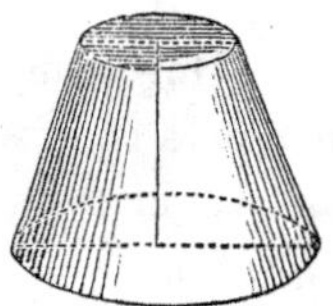

Operation.

We merely take the sum of the circumferences of the bases, and multiply by half the slant height, or side.

	8,4
	6
	14,4
half side	6,25
area	=90 *sq. ft.*

2. What is the entire surface of the frustum of a cone, the side being 16 feet, and the radii of the bases 2 and 3 feet?

Ans. 292,1688 *sq. ft.*

3. What is the convex surface of the frustum of a cone, the circumference of the greater base being 30 feet, and of the less 10 feet; the slant height being 20 feet?

Ans. 400 *sq. ft.*

4. Required the entire surface of the frustum of a cone whose slant height is 20 feet, and the diameters of the bases 8 and 4 feet *Ans.* 439,824 *sq. ft*

PROBLEM VI.

To find the solidity of the frustum of a cone.

RULE.

I. *Add together the areas of the two ends and a geometrical mean between them.*

II. *Multiply this sum by one-third of the altitude and the product will be the solidity.*

EXAMPLES.

1. How many cubic feet in the frustum of a cone whose altitude is 26 feet, and the diameters of the bases 22 and 18 feet?

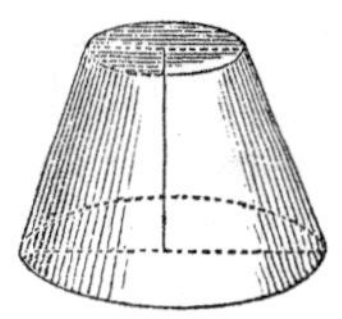

First, $\overline{22}^2 \times ,7854 = 380,134 =$ area of lower base:

and $\overline{18}^2 \times ,7854 = 254,47 =$ area of upper base.

Then, $\sqrt{380,134 \times 254,47} = 311,018 =$ mean.

Then, $(380,134 + 254,47 + 311,018) \times \frac{26}{3} = 8195,39$ which is the solidity.

2. How many cubic feet in a piece of round timber the diameter of the greater end being 18 inches, and that of the less 9 inches, and the length 14,25 feet? *Ans.* 14,68943.

3. What is the solidity of a frustum, the altitude being 18, he diameter of the lower base 8, and of the upper 4?

Ans. 527,7888.

4. If a cask, which is composed of two equal conic frustums joined together at their larger bases, have its bung diameter 28 inches, the head diameter 20 inches, and the length

40 inches, how many gallons of wine will it contain, there being 231 cubic inches in a gallon? *Ans.* 79,0613.

PROBLEM VII.

To find the surface of a sphere.

RULE.

Multiply the circumference of a great circle by the diameter, and the product will be the surface (Bk. VI. Th. xxiii).

EXAMPLES.

1. What is the surface of the sphere whose centre is C, the diameter being 7 feet?

Ans. 153,9384 *sq. ft.*

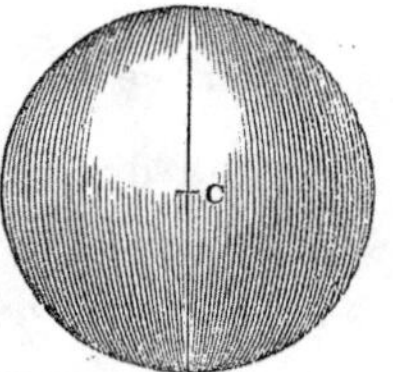

2. What is the surface of a sphere whose diameter is 24?

Ans. 1809,5616.

3. Required the surface of a sphere whose diameter is 7921 miles. *Ans.* 197111024 *sq. miles.*

4. What is the surface of a sphere the circumference of whose great circle is 78,54? *Ans.* 1963,5.

5. What is the surface of a sphere whose diameter is $1\frac{1}{3}$ feet? *Ans.* 5,58506 *sq. ft.*

PROBLEM VIII.

To find the convex surface of a spherical zone.

RULE.

Multiply the height of the zone by the circumference of a great circle of the sphere, and the product will be the convex surface (Bk. VI. Th. xxiv).

EXAMPLES.

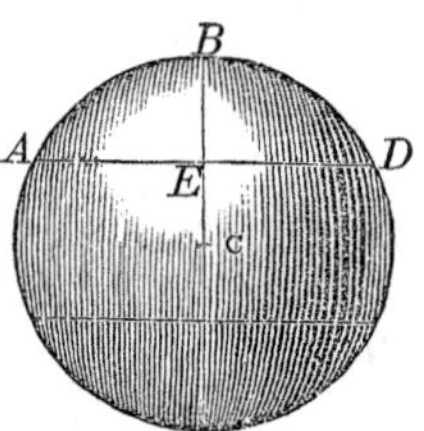

1. What is the convex surface of the zone ABD, the height BE being 9 inches, and the diameter of the sphere 42 inches?

First, $42 \times 3{,}1416 = 131{,}9472 =$ circumference.

height $= 9$

surface $= 1187{,}5248$ square inches.

2. The diameter of a sphere is $12\frac{1}{2}$ feet: what will be the surface of a zone whose altitude is 2 feet?

Ans. 78,54 *sq. ft.*

3. The diameter of a sphere is 21 inches: what is the surface of a zone whose height is $4\frac{1}{2}$ inches?

Ans. 296,8812 *sq. in.*

4. The diameter of a sphere is 25 feet and the height ot the zone 4 feet: what is the surface of the zone?

Ans. 314,16 *sq. ft.*

5. The diameter of a sphere is 9, and the height of a zone 3 feet: what is the surface of the zone?

Ans. 84,8232.

PROBLEM IX.

To find the solidity of a sphere.

RULE I.

Multiply the surface by one-third of the radius and the product will be the solidity (Bk. VI. Th. xxv).

EXAMPLES.

1. What is the solidity of a sphere whose diameter is 12 feet?

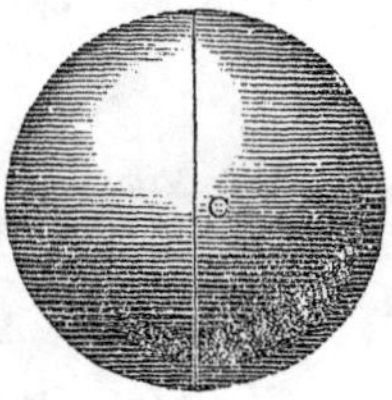

First, 3,1416 × 12 = 37,6992 = circumference of sphere.

diameter	=	12
surface	=	452,3904
one-third radius	=	2
Solidity	=	904,7808 cubic feet.

2. The diameter of a sphere is 7957,8: what is its solidity?
Ans. 263863122758,4778.

3. The diameter of a sphere is 24 yards: what is its solid contents?
Ans. 7238,2464 *cubic yds.*

4. The diameter of a sphere is 8: what is its solidity?
Ans. 268,0832.

5. The diameter of a sphere is 16: what is its solidity?
Ans. 2144,6656.

RULE II.

Cube the diameter and multiply the number thus found, by the decimal ,5236, and the product will be the solidity.

EXAMPLES.

1. What is the solidity of a sphere whose diameter is 20?
Ans. 4188,8.

2. What is the solidity of a sphere whose diameter is 6?
Ans. 113,0976.

3. What is the solidity of a sphere whose diameter is 10?
Ans. 523,6.

PROBLEM X.

To find the solidity of a spherical segment with one base.

RULE.

I. *To three times the square of the radius of the base, add the square of the height.*

II. *Multiply this sum by the height, and the product by the decimal* ,5236, *the result will be the solidity of the segment.*

EXAMPLES.

1. What is the solidity of the segment ABD, the height BE being 4 feet, and the diameter AD of the base being 14 feet?

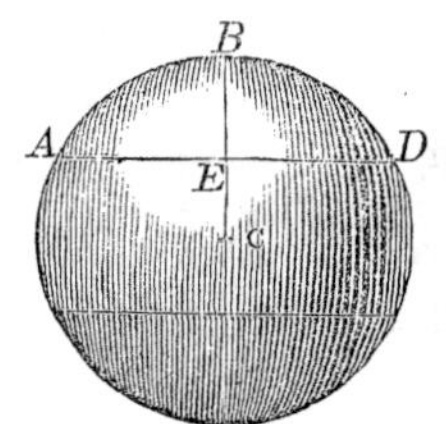

First,

$$\overline{7}^2 \times 3 + \overline{4}^2 = 147 + 16 = 163:$$

Then, $163 \times 4 \times ,5236 = 341,3872$ solid feet, which is the solidity of the segment.

2. What is the solidity of the segment of a sphere whose height is 4, and the radius of its base 8? *Ans.* 435,6352.

3. What is the solidity of a spherical segment, the diameter of its base being 17,23368, and its height 4,5?

Ans. 572,5566.

4. What is the solidity of a spherical segment, the diameter of the sphere being 8, and the height of the segment 2 feet? *Ans.* 41,888 *cubic ft.*

5. What is the solidity of a segment, when the diameter of the sphere is 20, and the altitude of the segment 9 feet?

Ans 1781,2872 *cubic ft.*

OF THE SPHEROID.

A spheroid is a solid described by the revolution of an ellipse about either of its axes.

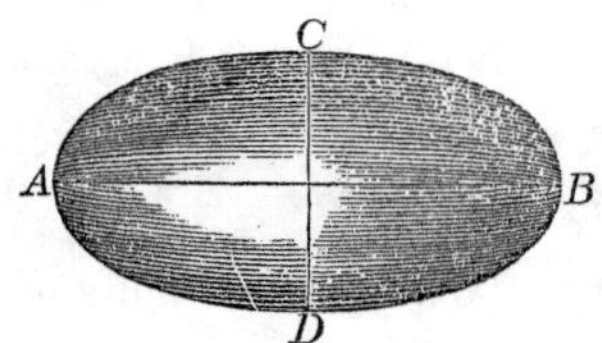

If an ellipse $ACBD$, be revolved about the transverse or longer axis AB, the solid described is called a *prolate spheroid:* and if it be revolved about the shorter axis CD, the solid described is called an *oblate spheroid.*

The earth is an oblate spheroid, the axis about which it revolves being about 34 miles shorter than the diameter perpendicular to it.

PROBLEM XI.

To find the solidity of an ellipsoid.

RULE.

Multiply the fixed axis by the square of the revolving axis, and the product by the decimal ,5236, *the result will be the required solidity.*

EXAMPLES.

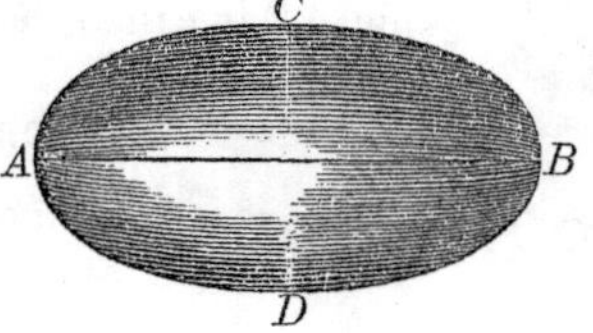

1. In the prolate spheroid $ACBD$, the transverse axis $AB=90$, and the revolving axis $CD=70$ feet: what is the solidity?

Here, $AB=90$ feet: $\overline{CD}^2=\overline{70}^2=4900$: hence

$AB\times\overline{CD}^2\times,5236=90\times4900\times,5236=230907,6$ cubic feet, which is the solidity.

2. What is the solidity of a prolate spheriod, whose fixed axis is 100, and revolving axis 6 feet? *Ans.* 1884,96.

3. What is the solidity of an oblate spheroid, whose fixed axis is 60, and revolving axis 100? *Ans.* 314160.

4. What is the solidity of a prolate spheroid, whose axes are 40 and 50? *Ans.* 41888.

5. What is the solidity of an oblate spheroid, whose axes are 20 and 10? *Ans.* 2094,4.

6. What is the solidity of a prolate spheroid, whose axes are 55 and 33? *Ans.* 31361,022.

7. What is the solidity of an oblate spheroid, whose axes are 85 and 75? *Ans.* ——

OF CYLINDRICAL RINGS.

A cylindrical ring is formed by bending a cylinder until the two ends meet each other. Thus, if a cylinder be bent round until the axis takes the position *mon*, a solid will be formed, which is called a cylindrical ring.

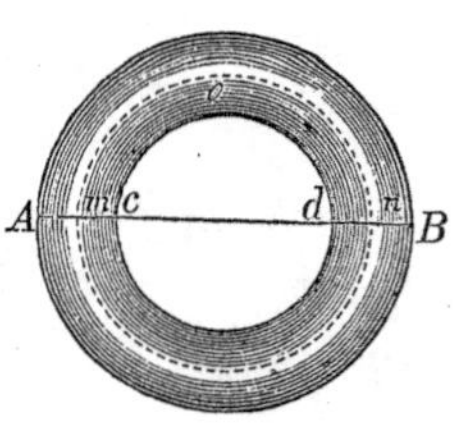

The line *AB* is called the outer, and *cd* the inner diameter

PROBLEM XII.

To find the convex surface of a cylindrical ring.

RULE.

I. *To the thickness of the ring add the inner diameter.*

II. *Multiply this sum by the thickness, and the product by* 9,8696, *the result will be the area.*

EXAMPLES.

1. The thickness Ac, of a cylindrical ring is 3 inches, and the inner diameter cd, is 12 inches: what is the convex surface?

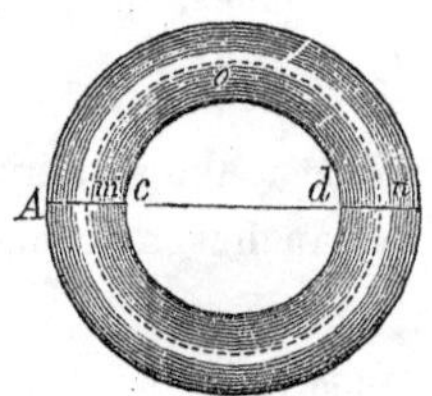

$Ac + cd = 3 + 12 = 15$:

$15 \times 3 \times 9{,}8696 = 444{,}132$ square inches = the surface.

2. The thickness of a cylindrical ring is 4 inches, and the inner diameter 18 inches: what is the convex surface?

Ans. 868,52 *sq. in.*

3. The thickness of a cylindrical ring is 2 inches, and the inner diameter 18 inches: what is the convex surface?

Ans. 394,784 *sq. in.*

PROBLEM XIII.

To find the solidity of a cylindrical ring.

RULE.

I. *To the thickness of a ring add the inner diameter.*

II. *Multiply this sum by the square of half the thickness, and the product by* 9,8696, *the result will be the required solidity.*

EXAMPLES.

1. What is the solidity of an anchor ring, whose inner diameter is 8 inches, and thickness in metal 3 inches?

$8 + 3 = 11$: then, $11 \times (\frac{3}{2})^2 \times 9{,}8696 = 244{,}2726$, which expresses the solidity in cubic inches.

2. The inner diameter of a cylindrical ring is 18 inches, and the thickness 4 inches: what is the solidity of the ring?

Ans. 868,5248 *cubic inches*

3. Required the solidity of a cylindrical ring whose thickness is 2 inches, and inner diameter 12 inches?

Ans. 138,1744 *cubic in.*

4. What is the solidity of a cylindrical ring, whose thickness is 4 inches, and inner diameter 16 inches?

Ans. 789,568 *cubic in.*

5. What is the solidity of a cylindrical ring, whose thickness is 8 inches, and inner diameter 20 inches?

Ans. ——

6. What is the solidity of a cylindrical ring whose thickness is 5 inches, and inner diameter 18 inches?

Ans. ——

THE END.

www.ingramcontent.com/pod-product-compliance
Lightning Source LLC
LaVergne TN
LVHW011209110826
845150LV00006B/1375

* 9 7 8 1 4 2 5 5 1 7 9 3 9 *